Strategic Positioning in the Oil Industry

STRATEGIC POSITIONING IN THE OIL INDUSTRY

Trends and Options

EDITED BY PAUL STEVENS

THE EMIRATES CENTER FOR STRATEGIC STUDIES AND RESEARCH

The Emirates Center for Strategic Studies and Research

The Emirates Center for Strategic Studies and Research (ECSSR), established on March 14, 1994 in Abu Dhabi, United Arab Emirates, is an independent research institution dedicated to the promotion of professional research and educational excellence in the United Arab Emirates (UAE) and the Gulf area. ECSSR serves as a focal point for scholarship on political, economic, and social issues pertinent to the UAE, the Gulf, and the Middle East through the sponsorship of empirical research and scientific studies conducted by scholars from around the globe.

The Center seeks to provide a forum for the scholarly exchange of ideas on these subjects by hosting conferences and symposia, organizing workshops, sponsoring a lecture series featuring prominent scholars and international dignitaries, and publishing original and translated books, research papers, studies, and two occasional-paper series. ECSSR also sponsors an active program that provides research fellowships, as well as grants for the writing of scholarly books and for the translation into Arabic of works relevant to the Center's mission. To support its research activities, ECSSR has acquired specialized holdings for its UAE Federation Library, and has created a state-of-the-art Information Center. Utilizing current Internet capabilities and on-line services, ECSSR is establishing an electronic database that will be a unique and comprehensive source for information on Gulf topics.

Through these and other activities, ECSSR aspires to engage in mutually beneficial professional endeavors with comparable institutions worldwide, and to contribute significantly to the general educational and scientific development of the UAE.

Published in 1998 by
The Emirates Center for Strategic Studies and Research
PO Box 4567, Abu Dhabi, United Arab Emirates
e-mail: root@ecssr.ac.ae
website: http://www.ecssr.ac.ae

Distributed by British Academic Press, an imprint of I.B.Tauris,
Victoria House, Bloomsbury Square, London WC1B 4DZ;
175 Fifth Avenue, New York, NY 10010

A full CIP record of this book is available from the British Library
A full CIP record of this book is available from the US Library of Congress

Library of Congress catalog card number: available

ISBN 1 86064 362 0 hardback
ISBN 1 86064 366 3 paperback

Contents

Figures and Tables

Figures

Tables

Abbreviations and Acronyms

ADNOC	Abu Dhabi National Oil Company
AGIP	Aziènda Generale Italiana Petroli
AL	Arabian Light Crude
APEC	Asia Pacific Economic Co-operation
Apicorp	Arab Petroleum Investments Corporation
ASEAN	Association of South-East Asian Nations
bl	barrel
b/cf	billion cubic feet
b/d	barrels per day
BHP	Broken Hill Proprietary Company Limited (Australia)
bn	billion
BNP	Banque Nationale de Paris
BOOT	build-own-operate-transfer
BP	British Petroleum
bp	basis price
BPCL	Bharat Petroleum
BZW	Barclays de Zoete Wedd
CIBC	Canadian Imperial Bank of Commerce
CIF	Customs-insurance-freight
CNPC	China National Petroleum Corporation
CO_2	carbon dioxide
CPC	China Petroleum Corporation
DEA	diethanolamine
Desul.	Desulfurization
DKB	Dai-Ichi Kangyo Bank (Japan)
DREAM	Domestic Regional Analysis of Margins (Salomon Brothers, New York)
dwt	dead-weight tons
ECA	Economic Cooperation Act
ECGD	Export Credits Guarantee Department
EGPC	Egyptian General Petroleum Corporation
ETBE	ethyl tertiary butyl ether

EU	European Union
EWC	East-West Center, University of Hawaii
EX IM	Export-Import Bank
FSU	former Soviet Union
g/l	grams per liter
GCC	Gulf Co-operation Council
GIC	Gulf Investment Corporation
gp	group
IBJ	Industrial Bank of Japan
IBK	Industrial Bank of Kuwait
IEA	International Energy Agency
IFC	International Finance Corporation
ING	Internationale Nederlands Groep
INOC	Iraqi National Oil Company
IOC	Indian Oil Corporation
IPG	Independent Petroleum Group
IPIC	International Petroleum Investment Corporation
IPO	initial public offering
IPP	Independent Power Project
KNPC	Kuwait National Petroleum Company
KPC	Kuwait Petroleum Company
KPI	Kuwait Petrochemical Industries
Kwh	kilowatt-hour
LDC	less developed countries
LIBOR	London Interbank Offered Rate
LNG	liquefied natural gas
LNOC	Libyan National Oil Corporation
m	million
M$	Malaysian dollar
MAGIC	Margin Analysis of Global International Countries (Salomon Brothers, New York)
mb/d	millions of barrels per day
MITI	Ministry of International Trade and Industry (Japan)
MPA	Maritime and Port Authority (Singapore)
MTBE	methyl tertiary butyl ether
mtoe	million tons of oil equivalent
MW	Megawatt

na	not applicable
NBC	National Broadcasting Corporation
NBK	National Bank of Kuwait
NIOC	National Iranian Oil Company
NNPC	Nigerian National Petroleum Corporation
NO	nitrous oxide
NOC	national oil company
O&GJ	*Oil and Gas Journal*
OECD	Organization for Economic Co-operation and Development
ÖMV	Östereichische Mineraloel Verwertung AG
ONGC	Oil and Natural Gas Corporation (India)
OOC	Oman Oil Company
OPEC	Organization of Petroleum Exporting Countries
PARCO	Pakistan-Arab Refining Company
PDO	Petroleum Development Oman
PDVSA	Petróleos de Venezuela SA
PEMEX	Petróleos Mexicanos
PIC	Petrochemical Industries Company
PIW	*Petroleum Intelligence Weekly*
psi	pounds per square inch
PSO	Pakistan State Oil
QGPC	Qatar General Petroleum Company
R/P ratio	Reserve to production ratio
RBS	Royal Bank of Scotland
RFG	reformulated gasoline
RON	research octane number
RVP	Reid vapor pressure
Saudi Aramco	Arabian-American Oil Company
SO	sulfur oxide
SRC	Singapore Refining Company
t/y	tons per year
tb/d	thousands of barrels per day
TFC	total final-energy consumption
toe	tons of oil equivalent
UAE	United Arab Emirates
UCC	Union Carbide Corporation
UK	United Kingdom

UNCTAD	United Nations Conference on Trade and Development
US	United States of America
USGC	United States Gulf Coast
USWC	United States West Coast
WTI	West Texas Intermediate
wts	measure of sulfur content by weight
Yen/Kl	yen per kiloliter
YPF	Yacimientos Petroliferos Fiscales

Note: not all abbreviations and acronyms used in this book are listed above

INTRODUCTION

Strategic Positioning in the Oil Industry: Trends and Options

Paul Stevens

The papers contained in this volume were among those presented to the Second Energy Conference held by the Emirates Center for Strategic Studies and Research, October 26-27, 1996. The First Energy Conference, held in October 1995, covered current and future trends in the world's energy markets. In particular, it looked at near-term United Arab Emirates (UAE) and Gulf Co-operation Council (GCC) energy market choices and constraints. The second conference took a narrower perspective, and focused on the strategic options open to various players in the light of the current oil industry and market context. That context is created by certain characteristics and trends. These characteristics are to a large extent immutable, while the trends are constantly changing. A brief discussion of both will set the context for the chapters which follow.

Characteristics of the Industry That Set the Context for Strategic Decisions

FEAST OR FAMINE?

From the very early years, the stages of the value chain which constitute the international oil industry have been characterized by feast or famine.[1] Each stage, at different times and in different contexts, has suffered from shortage or excess capacity. If the last quarter century is taken, then the evidence is compelling.

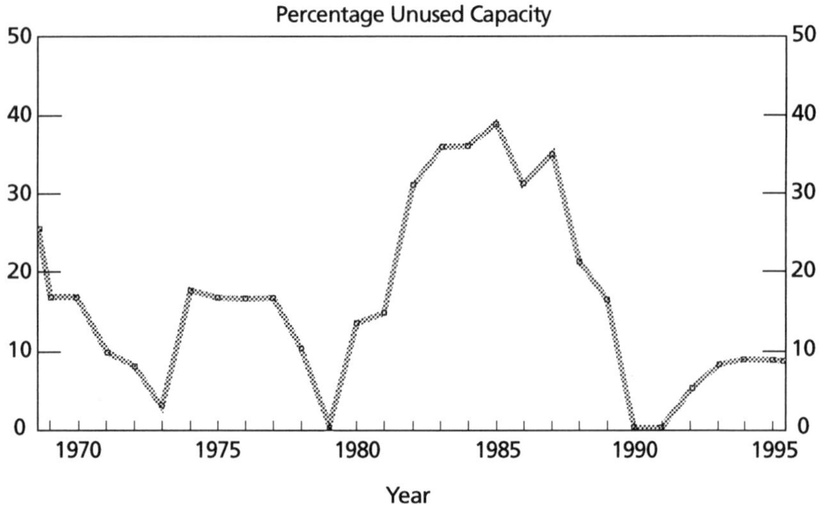

Source: Capacity-CIA, Production, BP

Fig. Intro. 1 OPEC Excess Capacity to Produce Crude Oil 1968-1996

In the production of crude oil, as can be seen from Fig. Intro. 1, the overwhelming tendency has been for excess capacity to reside with OPEC (Organization of Petroleum Exporting Countries),[2] with only three periods when the excess has been eroded.[3] The first period was in the early 1970s. Expectations of lower prices in the late 1960s led to purchases of oil-fired equipment, creating dramatic demand growth up to 1973. The announcement of "participation" as an aim by OPEC, coupled with the consequent negotiations, mitigated against the development of new capacity. Into this situation of strong demand and stagnant capacity came the Arab oil embargo associated with the 1973 Yom Kippur War, which further aggravated the situation. The second period of no-excess capacity was associated with the Iranian oil workers' strike of 1978, which was followed by the Iranian Revolution. Although the actual strike lasted for a short period, the new revolutionary government, invoking intergenerational concerns, imposed an effective ceiling on future production. This erosion of excess capacity was reinforced by Iraq's invasion of Iran in

September 1980. The most recent period when excess capacity disappeared came as the result of UN sanctions following Iraq's invasion of Kuwait in 1990.

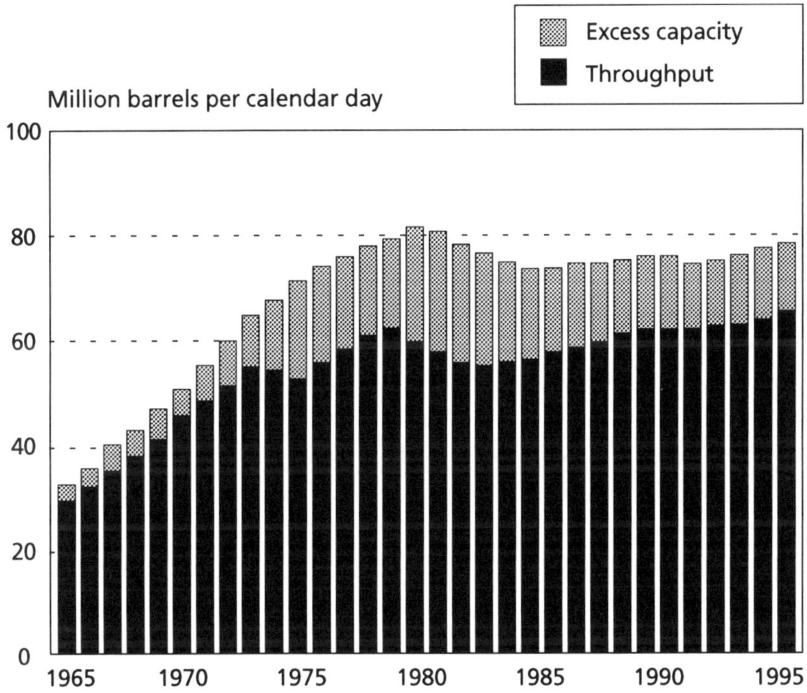

Source: BP, 1997

Fig. Intro. 2 World Refinery Capacity and Throughput 1965-1996

Refining has also tended to experience more feast than famine in recent years.[4] During the 1960s, refinery capacity managed to keep pace with the dramatic increase in oil demand driven by the OECD's (Organization for Economic Co-operation and Development) economic miracle triggered by the Marshall Plan. This is reflected in Fig. Intro. 2.[5] However, towards the end of the 1960s, expectations of ever-rising demand persuaded the oil companies to increase their refinery-building program in anticipation of growing demand. Just as the capacity began to come onstream, the expected demand growth was frustrated by the first two oil shocks of 1973 and 1979–81. The result, as can be seen from Fig. Intro. 2, was very large amounts of excess refinery capacity. The situation was compounded because the lead-times on refinery projects meant that new capacity emerged long after the discontinuity generated by 1973. A similar story can be told for refineries on a more regionalized level.[6] For example, in recent years the Atlantic Basin has consistently suffered from an excess capacity to produce gasoline, which stems from the impact of reformulated gasoline capacity forced by the US Clean Air Act. Another example has emerged in Asia. In the late 1980s, there was a growing view that the demand barrel, especially in Asia, was getting lighter, while the expectation was that the crude slate would get heavier. As a result, refiners began to invest heavily in upgrading capacity, especially in Asia. However, by the time the kit came onstream, while the demand barrel had lightened, the crude slate had not become heavier. The result was that the crude price differentials made much of the investment in upgrading look less than attractive.

In recent years, as can be seen from Fig. Intro. 2, excess refinery capacity has been reduced through a combination of scrapping capacity and the post-1986 recovery in oil demand. As will be seen below, generally poor refinery profitability has caused some to be concerned that insufficient investment in new capacity may eventually create shortages. The problem is that while this may presage more favorable refinery margins, potential investors still bear the scars of the last time they responded to the obvious signals in the market. As a result, investment decisions may well be inhibited.

A similar story can be told of the tanker market.[7] As with refineries, expectations of ever-growing demand in the late 1960s persuaded tanker owners that more tonnage would be required. Furthermore, the pressure was to build much larger tankers. This was driven by two factors. The closure of the Suez Canal in June 1967 removed previous size constraints, and the enforced longer voyage around Africa gave full scope to the benefits of economies of scale. Second, technological developments in shipbuilding allowed tankers to take full advantage of their potential for huge economies of scale. However, just as the new capacity began to appear in any volume, the first oil shock reversed the demand trend, and the tanker market suddenly found itself with very significant excess capacity. Over the years, this excess has been eroded. In the meantime, little new capacity has been added, and the fleet is aging. By 1995, of the fleet of 443 tankers over 200,000dwt, the workhorses of the ocean-going, crude-carrying capacity, 263 were built before 1980. This means that by the end of the decade, they will have reached the end of their design life.[8] This again has raised concerns of impending shortage, but as with refining, previous scars still run deep, and there is little sign of much investment in new tanker capacity.

The Reasons for Feast or Famine

A number of factors help explain why the industry has this tendency to over- or under-supply. Projects at all stages tend to have long lead times, which means that by the time the investment decision has converted into actual capacity, the world may well have changed. The long lead times arise from the fact that the projects are invariably highly capital intensive, and are often technically complex, requiring sophisticated engineering. Care and capital equipment demand time. Also, typically, projects in the oil industry tend to be large-scale. Because oil flows in three-dimensional space, it is subject to very large technical economies of scale. In tanks or pipes, which are the equipment of the industry, capital costs are a function of surface area while output is a function of volume. There exists an exponential relationship between the two. For example, a pipeline's capacity equals the square of its radius. Hence, large-scale projects tend to attract much lower average costs. In the oil industry, big is beautiful.

Coupled with long lead times is the fact that the industry has always been characterized by a strong tendency for everyone to make the same investment decisions at the same time, en masse. This can be explained by several factors. Investors tend to respond to the same signals. If the signals appear to be pointing clearly in one direction, then the decision appears inevitable. In addition, there exists a phenomenon that can be called "forecasters' cluster." Corporate planners in oil companies all tend to read the same trade press and attend the same conferences. Increasingly, as part of the recent process of downsizing corporate headquarters, they are also outsourcing to the same forecasting companies. The result is constant exposure to the same ideas and views on the future. Furthermore, there is safety in numbers. To go with a consensus view of the future provides safety: if you are wrong, so is everyone else. To go counter to the trend is a high-risk strategy. This is reinforced because often in the oil industry, at least in the early stages, consensus tends to produce behavior that is self-fulfilling. It is only after some time — reflecting the long lead times discussed above — that the consensus forecast inevitably sows the seeds of its own destruction.

The Implications of Feast or Famine

From the fact of feast and famine, there follow a number of crucial implications that set the context in which players in the industry must operate and make strategic decisions.

The cost structure of most oil-industry operations, with their very heavy weighting towards fixed or sunk costs, means that the "bygones rule" is supreme. Hence, when an operator is considering whether to continue an unprofitable activity, the key question is whether variable costs are being covered. If so, since fixed costs have to be paid anyway (bygones are bygones), losses are minimized by continued operation. This explains why the industry can face feast without total collapse. Hence, while tanker owners and refiners were making significant short-run losses in the worst days of over-supply, it was still worth maintaining operations. An important implication for the operation of the industry is that the distortions resulting from faulty investment decisions remain dominant in the industry for a long time after their occurrence, reflecting both lead times and the relatively long life of equipment. Eventually, however, the short-term moves into the longer-term. This requires decisions to convert the fixed costs into variable costs, either by scrapping or by selling the equipment. It is during such periods of rationalization that strategic opportunities occur, either to divest or to acquire capacity. Such reshuffling of assets is often at the core of industry restructuring.

A second implication of feast or famine is that companies are encouraged to hedge against the risk of decisions in one sector being made poor decisions by changed circumstances that result from successful decisions in another part of the value chain. Hence, positioning in all stages of the value chain may provide some protection through a process of diversification. This is reinforced, since the "price" of one stage of the chain represents the "costs" of the next step in the chain. In rather simplistic terms, for example, excess refinery capacity may lead to lower refinery gate-prices, greater volume of output, and hence stronger demand for crude. Similarly, excess crude availability lowering crude prices may well increase refinery margins.

A third implication is that the industry periodically faces the problem that it is living on past investments, and must decide how to position itself for the future. Just as the state of capacity appears to be cyclical, so is industry restructuring. However, such positioning is always in danger of being paralyzed by the uncertainty of rivals' behavior, the classic oligopoly problem in economics. Hence, if your rivals invest as you do, feast may follow, and with it excess capacity and losses. If others do not invest, creating famine, and you do invest, you then stand to gain from access to capacity in a scarce market. However, as this uncertainty moves into second-order thinking, it can get extremely complex, and may well generate a paralysis over investment.

THE IMPERATIVE OF CAPACITY OPERATION

A second immutable characteristic that can be noticed in the international industry is the need to operate each piece of equipment and each system as close to full capacity as possible. Much of the history of Standard Oil in the US during the last century can be written in terms of ensuring that the Rockefeller pipelines operated at full capacity.[9] Much of the poor profitability record associated with refineries following 1973 can be explained by the below-capacity operation seen in Fig. Intro. 2.[10]

The Reasons Requiring Capacity Operation

The reasons lie with the cost structure associated with each stage in the value chain. With very large fixed costs, below-capacity operation means that fixed costs are spread over smaller throughput, causing average fixed costs to rise exponentially. Such rises damage profits.

The Implications of the Imperative for Capacity Operation

The key implication of this simple fact concerns security of supply on the input side and security of market on the output side. Much of the history of the industry concerns the way companies have positioned themselves to ensure such security. Problems associated with the writing and enforcing of contracts are legion. Long-term contracts face the problem of changing circumstances, which may convert a good contract into a bad contract, which then continues to rule the relationship. Short-term contracts that avoid such inflexibility and vulnerability to change of course lose the imperative of supply security. It was such problems with contracts that in part drove the vertical integration of the international oil industry, ensuring a company's presence in all stages of the value chain.

At this stage, definitions of vertical integration are required. Financial vertical integration is when a company owns (i.e. controls the cashflow of) affiliates in different stages of the industry. Operational vertical integration occurs when one affiliate explicitly supplies its output as the input to the next affiliate down the chain. Clearly, financial vertical integration is a prerequisite if operational vertical integration is to be present. However, operational vertical integration is not an automatic consequence of financial vertical integration. Instead, the affiliates may simply choose to use markets. Hence, the crude-producing affiliate sells into the crude oil market, the refining affiliate buys from the crude market and sells into the product market, and finally, the marketing and distribution affiliates buy from the product market.

Until the 1970s, the major oil companies were both financially and operationally vertically integrated. This had an added bonus of providing very significant barriers to entry, thereby restricting competition. To be a refiner required a crude slate. If all the crude was moving on an inter-affiliate basis,[11] then no crude slate was available, unless secured directly by the would-be refiner. Operational vertical integration also enabled price discrimination, whereby the company could vertically integrate into the demand-elastic (hence low-price) market to prevent resale into the inelastic (high-price) market. To some extent, the process encouraging operational vertical integration was self-feeding. Widespread operational vertical integration meant very limited arms-length offering. This made the crude market small and inefficient, generating high transaction costs if the market was used. In turn, this gave ever-greater logic to being vertically operational, further reducing arms-length offering, reducing market size and efficiency, and increasing transactions costs.

The situation was altered by the take-over of the joint-venture operating companies by the producing governments between 1972 and 1976. This did not immediately cause a change since, in many cases, the producer governments allowed the previously owning companies to retain their preferential role in marketing crude oil. Only with the breaking of long-term sales contracts by the producer governments during the second oil shock of 1979–81 did operational vertical integration begin to disappear.[12] As operational vertical integration declined, the self-feeding process began to reverse itself. More arms-length offering from the producing governments increased the volume of transactions in the market, and with it market efficiency. As market-transaction costs fell, private companies began to move voluntarily away from operational vertical integration, thereby further increasing market volumes. By the 1990s, very little of the privately owned oil company operations could be described as operationally vertically integrated.

However, at the same time, the oil-producing governments have developed a downstream capability, in part to guarantee security of markets. Where they have done so, they behave on an operationally vertically integrated basis. An issue of key strategic importance facing these companies, as Luciani and Salustri point out in Chapter 1, is how far they will continue this process, since it carries important implications for the volume of offerings in the various markets.

THE MOTIVATIONS OF PLAYERS CHANGE

Three broad classes of player in the international oil industry can be identified. The first are privately owned companies whose shares can be bought and sold on the relevant stock exchange. These are predominantly financially vertically integrated companies, although recent years have seen the rise of a substantial number of relatively small upstream companies with no downstream capabilities. This group of privately owned companies is still dominated by the major oil companies, the Seven Sisters, of which only Gulf Oil, taken over by Chevron, failed to survive. Second, there are the national oil companies (NOCs) of the oil-producing countries. These are state-owned and -controlled companies, with varying degrees of both financial and operational vertical integration. There have been some cases in recent years, such as Yacimientos Petroliferos Fiscales (YPF) in Argentina and Petronas of Malaysia, where the drive for privatization has led to a company moving from this group into the first group of private companies. Third, there are companies that effectively became the national champions of the consuming countries. These companies were largely state-owned and -controlled, although, as with the NOCs, they too have been moving into the category of private companies as privatization has taken place.

There is a relevant further distinction. Private companies quoted on stock exchanges in North America or the United Kingdom — the so-called Anglo-Saxon model — face a rather different set of imperatives to those quoted on stock exchanges following what has been termed the German-Japanese model. At the risk of oversimplifying what is extremely complex,[13] the Anglo-Saxon model is where fund managers who dominate the markets seek to maximize return, and where poor performance tends to produce adverse shareholder behavior, i.e. selling of shares, which compounds the effects of poor share performance. By contrast, the German-Japanese model is characterized by having as dominant shareholders large commercial banks, which tend to be involved in the management of companies they control by virtue of their shareholding. The distinction has led some observers to suggest that the Anglo-Saxon model has a strong short-term orientation that forces attention only as far as the next quarter's financial performance, while the German-Japanese model is claimed to have greater emphasis on the long-term operating performance. The distinction is potentially important for the oil industry, which is characterized by the need for relatively long-term investments.

The motivations of these three categories of player differs on both a cross-section basis (i.e. between them) and a time-series basis. Changes over time in the motivation of the private companies have been well documented.[14] While the constant theme has been something called "profit maximization," how this is defined in an operational sense changes. For example, during the second oil shock of 1979–81, most private companies had an extremely vague notion of profit maximization. Hence, when many gained huge windfall profits as a result of the much higher oil prices, many companies spent it on a process of diversification, not only into other areas of energy, but also on areas totally unconnected with the industry.

Currently, as a result of developments in the theory of financial economics, profit maximization for companies operating in the Anglo-Saxon model is defined under the rubric of "value-based management" as maximizing shareholder value. This means the company should provide a return to the shareholder at least on a par with other companies in the same sector. If for any reasons the company is unable to achieve this, the company should return money to the shareholders, and allow them to create their own value added by investing elsewhere.

What the NOCs of both producer and consumer governments pursue has also changed. In the early days, producer-government companies sought control over the country's resources. As this control became a reality, other objectives began to emerge, including the promotion of economic development and securing market outlets. Consumer-government companies were generally motivated by a desire to secure crude oil access for national strategic reasons, and in some cases, to provide a counterweight to the downstream market power of the Seven Sisters. In all cases, in the nature of such institutions, there also arose a dichotomy between the objectives of government, as espoused by politicians, and the rent-seeking behavior of the company technostructure. Probably the most extreme example of this was provided by Pemex in Mexico, which virtually became a state within a state. Many countries experienced something similar, if not quite on the same scale.

The Reasons Motivations Change

Two reasons explain the differing and changing motivations. The first concerns the intellectual output from the universities and business schools, which provides an ideological basis which drives motives. Thus, "value-based management" came out of theories associated with portfolio analysis and the capital asset-pricing model. Similarly, awareness and concern over the behavior of the NOC bureaucracies emerged from the development of theories of public choice, and later principal-agent analysis. All of these provided a set of ideas that has permeated corporate America, and from there trickled down to other regions. In the early 1990s, such ideas encouraged a trend of "re-engineering" and "right-sizing." More recently, this has switched into seeking growth in new markets, consolidation in mature markets, and emphasis on competitive advantage. In this sense, the oil companies have been little different from other companies in following such behavior. The only difference is that the ability of oil companies to attract better-quality manpower, more likely to be aware of changes in the intellectual underpinning of many of the managerial ideas, speeds their transmission throughout the corporations.

The second driver of change in motivation has been ownership changes. The first big change came in the 1970s, when governments took over the operating companies through a combination of straight nationalization and creeping participation. More recently, a new period of changing ownership is emerging, associated with deregulation and privatization. This will be discussed in more detail below.

The Implications of Changing Motivations

The implications of differing motivations depends upon which stage in the cycle is being considered. The current state of affairs will be considered in the next section. However, the most obvious implication in general terms is that not only is the operating context changing, but the response of competing players is also liable to change. This therefore requires both tactical and strategic flexibility and imagination for a company to be successful. It also has significant implications for the nature of the competitive environment and the time horizons of the various players.

The Current Trends That Set the Context for Strategic Decisions

Many different trends are currently affecting the operating context, and hence the strategic positioning of oil companies. This introduction will consider three trends: moves toward privatization and deregulation; greater revenue volatility; and the growth of environmental concern.

PRIVATIZATION AND DEREGULATION

Driving Forces

The oil industry in recent years has been experiencing both deregulation in previously highly regulated markets, such as Japan, and privatization of state-owned oil companies, both "producers," such as YPF and Petronas, and "consumers," such as ENI and Total.[15] Privatization and deregulation in the oil sector have been part of the general trend favoring such policies. Several factors explain this.[16] During the 1980s, a whole set of intellectual arguments began to permeate the policy making process. Thus, ideas associated with the economic theory of politics and theories of public choice began to emphasize the negative impact of state intervention, and promote greater reliance on the use of markets. This process was reinforced as the attention of OECD policy makers focused away from unemployment and Keynesian-type intervention to inflation and monetarist laissez-faire. With missionary zeal, such views were forced into the policy debate by the International Monetary Fund (IMF) and the World Bank, the so-called "Washington Consensus." The debt crisis and the enforced requirement for rescheduling provided the levers by which such views were imposed. However, it was increasingly the young technocrats in developing countries who willingly embraced the views that they had so recently been exposed to in Western universities. Finally, the collapse of the Soviet Union seemed to provide living proof of the negative impact of large-scale state involvement.

In addition to this general move in all sectors of the economy, there were also specific issues within the oil sector. The lower world oil price post-1986 forced the private companies into ever-greater concern with profitability. This was reinforced because their main areas of operation were the mature markets, where there existed much over-capacity, barriers to exit, greater competition, threatening margins, and in some cases, saturation of oil-product markets. As a result, many areas of oil operations began to experience significant cost reductions. Nowhere has this been more obvious than in offshore oil production.[17] A whole series of technological improvements — ranging from horizontal drilling, through subsea completion and offshore loading to three-dimensional seismic and many others — caused dramatic reductions in cost. Despite this trend of cost reduction for the private companies, it seemed to many that their own NOCs were becoming ever higher-cost, reflecting inefficiency and often corruption. Observers in many countries began to express concern about the "rent-seeking" behavior of the bureaucrats running the companies. This helped feed the growing pressures for privatization, which was seen in some cases as a panacea for the problems of the oil sector.

Issues

This context of the retreat of the state and the rise of markets gave rise to a great many issues relevant to strategic positioning. Firstly, it provided opportunities. Market access that had previously been denied suddenly became feasible under deregulation. However, national and political barriers survived to varying degrees, encouraging strategic alliances or joint ventures with local entities that can navigate such barriers. In many places, this is also giving a regional dimension to such alliances.

Secondly, companies that had previously been constrained in terms of where they could invest and through what mechanisms now faced new opportunities. The result of greater access and the means to benefit from such access meant an increase in the number of players at various stages in the industry, hence greater competition. While Luciani and Salustri see this as an irreversible process, it is possible to argue otherwise. A collapse in the crude oil price could introduce a period of consolidation whereby some of the smaller players are swallowed, or whereby certain geographic locations of the big players are sold in an effort to rationalize.

Greater competition arising from deregulation provides both threats and opportunities for existing companies. Competition forces low-cost operation, the use of opportunities presented by economies of scale, and the seeking of new opportunities to create niche markets. At the same time, privatization changed the motivations and objectives of previously state-owned companies. Profit became more of a driving force. These conditions persuaded many companies to consider strategic alliances, either to protect themselves from greater competition or to strengthen the company's ability to take advantage of opportunities. As well as linkages between oil companies, there is also growing interest in gas-oil alliances as gas is increasingly seen as the fuel of the future. Also, as Al-Moneef points out in Chapter 2, a similar trend creating links with the electricity sector is emerging. Deregulation is also opening up financing opportunities previously denied many of the state-owned players. In addition, the liberalization of financial markets, coupled with low real interest rates and high equity values, is making strategic moves and plans easier to finance.

REVENUE VOLATILITY

Driving Forces

The post-1986 oil markets were characterized by increasing volatility in cashflows associated with producing, refining, and marketing of crude oil.[18] Several factors contributed. First, the market controllers after 1986 surrendered their price-making role to the whims of competitive markets.

Although they still had to control excess capacity by matching crude supply to demand, once this range was made, price was determined by a growing number of players in an increasingly transparent market. The problem was that within a very wide range of prices, any price would clear the market. Therefore, the price was determined not at any equilibrium point of balanced supply and demand, but by the belief of those trading, who were in the business of anticipating what the average opinion of the average opinion might be. Furthermore, as the crude oil markets began to develop the type of derivative instrument long common in other commodity markets, speculators began to play a greater role. Despite much of the earlier conventional wisdom of economic theory, such speculation does increase volatility. This previously unfashionable view is now being supported by theoretical work associated with herd behavior and bubbles.[19] Increasingly, oil prices were set by belief, and belief can be ad hoc and whimsical, leading to significant price volatility.[20]

In addition to price volatility in the crude market, greater competition in the product market also put product prices under pressure. In particular, the entry of hypermarkets in Western Europe and mass merchandisers in the US as gasoline retailers had a significant impact on gasoline prices. Price volatility was also strengthened by increasing regionalization of oil markets. In earlier times, the oil market was most definitely global.[21] While it still remains so, two factors are pressing for more regionalized markets. Firstly, the period of extremely low transport costs resulting from the feast of tankers during the 1970s and 1980s is coming to an end. The significance is obvious. Regional price differentials can only trigger arbitrage if the transport costs are below the price differential. Higher transport costs will require greater spreads to induce a supply response. Secondly, greater variation in product specification and standards driven by environmental concerns also fragments the global product markets. The most striking example is the way in which the requirement for reformulated gasoline in the US has restricted the ability of that market to draw in supplies from sources without such standards.

For some oil producers, revenue volatility derived from oil prices has been compounded by volume fluctuation. In particular, it has been a longstanding complaint of OPEC members that their market share is threatened by the strength of non-OPEC supplies. More recently,[22] the problem is compounded by the fact that a number of OPEC countries, as part of the general deregulation process, have allowed the entry of foreign companies into their upstream. Their consequent ignoring of quotas has now provided a further source of concern over market share for those that remain true to quota in their market-control role, namely Saudi Arabia, Kuwait, and the UAE.

Market share issues within OPEC — and by implication, revenue volatility issues — have also been compounded by the Iraqi invasion of Kuwait. Firstly, there was the issue of who benefitted in volume terms from the loss of Iraqi and Kuwaiti crude. Secondly, there was the issue of market share and positioning for the eventual return of Iraq, and the implicit negotiations over quota that would be required by a full lifting of sanctions.

Issues

The growing revenue volatility raises a number of issues of a strategic nature. For one, volatility encouraged moves to hedge against the risk of a weak cashflow performance in one sector by a presence in another part of the value chain. Thus, weak crude revenue could be protected by higher refinery margins, while weak refinery margins could be offset by crude-price strength. There is also the argument emphasized by Luciani and Salustri that presence in both the crude-producing stage and the refining stage reduces the importance of crude prices, and gives producers greater control over their revenue. A variant on this theme is that downstream access locks in volume, thereby reducing revenue volatility from fluctuating market share.

ENVIRONMENTAL PRESSURE

Driving Forces

The final trend to be considered in this introduction relates to the rise of environmental concerns. This is a relatively recent phenomenon in relation to energy, taking place as climate change, acid rain, and ozone depletion come to the attention of the public.[23]

A further dimension, as indicated by Doshi in Chapter 4, regards its importance in the policy process. Many assumed that environmental awareness was a luxury good of real practical concern only to the relatively affluent. As a result, it was assumed that it would not drive policy in the developing world. However, such a view neglects the disproportionate importance of urban middle-income groups on policy making. In many developing countries, environmental issues, especially urban traffic pollution, are beginning to influence the policy debate. Hence, despite the expectations of many, environmental concerns are creating a global context influencing strategic decisions.

Issues

Environmental concern has many strategic dimensions. There is the impact on the location of investment decisions. In the days of operational vertical integration and its consequent transfer pricing, the major companies' objective was the minimization of the global tax bill. Today, there is a view that the large multinationals will seek to invest such that they can minimize the costs associated with environmental legislation. This is either because the legislation itself is weak or because existing legislation is not enforced. Of course, environmental legislation can also provide niche opportunities. In the US, the major oil companies initially resisted the Clean Air Act, until they realized that as the only players with sufficient resources to produce reformulated gasoline, they had effectively been presented with niche-market opportunities. A further consequence of such legislation is the fragmentation of global product markets referred to earlier, which also raises strategic issues.

Environmental legislation raises issues of interfuel substitution that have strategic implications. Natural gas, which has always been a fringe activity of the international oil industry, is moving center-stage as a green fuel providing significant opportunities for those positioned to take advantage. Finally, environmental legislation can rapidly turn assets into liabilities. An excellent example is the fact that such legislation has made it prohibitively expensive to exit from refining in much of the OECD. Such a threat may well persuade owners to shed these activities before the liability stage emerges. It is such considerations that are encouraging a process of downstream rationalization by the major private companies, creating possible downstream opportunities. However, at the same time, they are also generating reluctance by the potential buyers of such assets, who fear paying for what will soon become liabilities.

It is within the general context discussed in the previous two sections that the issues addressed by the five chapters in this volume are relevant.

The first two chapters, by Luciani and Salustri, and by Al-Moneef, address the key issue of vertical integration and investment by the NOCs into the downstream. Luciani and Salustri's premise is that the industry is currently experiencing a phase of reintegration, and they argue that this is part of a cycle observable over time in the industry, where fragmentation gives way to tight vertical integration, followed by a return to fragmentation. They argue that there is currently an increase in the number of players in the market, a trend that they view as irreversible. The chapter then attempts to measure the degree of integration by the computation of an "imbalance index." The results are mixed, suggesting that recent years have seen some forces that increase balance between the upstream and downstream and some that decrease the balance. In this context, Luciani and Salustri examine what the NOCs of the producer governments have been doing. They ask why the industry is not experiencing more vertical integration, and consider why NOCs are hesitating to invest more in the downstream. The chapter concludes with a strong argument that NOCs should vertically integrate because of their increasing control over crude prices, and to reduce their dependence on the whimsical movements of a small number of highly unrepresentative crudes that currently dominate term-contract pricing.

Al-Moneef provides a complementary chapter, which also looks at the moves by NOCs into the downstream in recent years. Al-Moneef's starting premise is that vertical integration is the essence of the industry, and argues that the subsequent diminution of such integration in the 1970s and 1980s has contributed significantly to the growing instability of the market, already commented on above in the section entitled "Revenue Volatility." He then considers in some detail the attempts by the NOCs to reintegrate by acquisition of downstream assets. Given this context, he considers the arguments for and against such downstream integration. In the arguments in favor, he stresses the synergies associated with joint ventures and strategic alliances. As for the arguments against, these revolve around the relatively poor performance of the downstream profitability compared to upstream rents or, indeed, investments in other non-related areas. He then turns to consider how these patterns may continue into the future, and the context in which these future decisions will be made. On balance, the chapter concludes by stating that the benefits of downstream involvement outweigh the costs.

The next two chapters, by Yokobori and Doshi, both introduce a regional dimension to the strategy decision by considering the situation in Asia. This emphasis on Asia needs little justification. In the years 1980–96, while world oil demand grew on average at 0.83 percent per year, Asia–Pacific oil demand grew at 3.65 percent, and the six Asian Tigers grew at 7.7 percent.[24] Most forecasts see future oil-demand growth predominantly in the Asia region. Yokobori's chapter begins with a detailed analysis of this emerging oil market, examining the main driver of economic growth and the implications for import dependence on the Gulf. This then provides a context in which the opportunities for Gulf oil companies are enumerated. A number of challenges to these companies are then considered. These include environmental concerns, security of supply, and finally, the price at which crude oil will be traded. Yokobori concludes by arguing that a mutuality of interest should be the basis for future developments.

The second paper with an Asian theme, by Doshi, starts with an overview of the world oil market. Out of this evaluation, then, comes a detailed discussion of product specifications in the Asia–Pacific region. The chapter makes the important point that environmental concerns in Asia are growing rapidly, driven by an increasingly vociferous, urban middle class. However, the result is an extremely heterogeneous set of specifications, reflecting differing interests and differing levels of development. Doshi details the specification for gasoline, diesel/gasoil, and fuel oil, and then describes how refiners in the region have responded to these changes in terms of sulfur and upgrading facilities. The chapter concludes by putting likely developments in Asian refining into a more global context.

The final chapter in the volume, by Al-Mazeedi, addresses the issue of privatization and deregulation. In particular, he looks at what has been happening in the Gulf. The argument is that full-scale privatization in the oil sector has not been possible, and instead there have been what the author describes as a series of "back-door" privatizations. Extensive details are given on what has taken place in the Gulf and elsewhere. Partial privatizations, it is argued, have resulted in a failure to gain the real benefits of privatization, with the result that the development of Gulf NOCs is being stunted. Specifically, partial privatization is frustrating the opportunity to convert comparative advantage in oil into longer-term competitive advantage. Al-Mazeedi argues that this is because the technostructure of the state-owned companies does not react as speedily as a private-sector management team would.

The reader should note that with the exception of the introduction and Majid Al-Moneef's chapter, the views and opinions expressed in this volume are not those of academics but of practitioners. Indeed, the essays contained herein offer an insight into the beliefs and perceptions of those actively involved in today's oil industry. Overall, the chapters confirm that the industry is responding to current pressures and trends, and that as a result, significant changes are underway. What remains to be seen, however, is how far the private and national oil companies will actually position themselves strategically in relation to both the current context and to each other.

I am particularly grateful to Dave Pender of the Emirates Center for Strategic Studies and Research, who has applied his own considerable editorial skills to the production of this book.

CHAPTER 1

Vertical Integration as a Strategy for Oil Security

Giacomo Luciani and Mario Salustri

Introduction

Is the international oil industry in a phase of reintegration? Answering this question is the main concern of this chapter, one that is justified by historical observation, which indicates that the industry has experienced phases of intense competition and fragmentation — with numerous independent operators engaging in each of its fundamental operations — as well as phases of tight vertical integration, in which a limited number of large operators engaged in all stages of the industry and internalized transactions from one stage to the next.

In a nutshell, the industry began in a state of extreme fragmentation, which led to extreme instability in the price of oil in the relevant markets, i.e. in the United States. Instability led to the ruin and insolvency of most operators, creating conditions facilitating the formation of the Standard Oil Trust, which represented the opposite extreme of acute monopolization. Following the break up of Standard and the formation of two major oil companies outside the US, the industry entered into a protracted phase of vertically integrated oligopolies, popularly known as the Seven Sisters. This phase ended in the early 1970s, when key producing countries nationalized the upstream properties of the major international oil companies. Although some of the companies maintained preferential access to crude oil, this proved to be a temporary privilege which disappeared with the second oil shock of 1979–81.

Two major effects ensued: firstly, a much larger share of global crude production came to be exchanged between independent entities; and secondly, some of the former major international oil companies either disappeared (such was the fate of Gulf Oil) or were seriously reduced in size.

The number of relevant international operators has therefore considerably grown. This is likely to be an irreversible trend. Indeed, one can say that it existed well before the 1970s, when independent operators — such as Occidental, Conoco and Hunt for example — and some state-owned companies — such as Italy's ENI — slowly loosened the Seven Sisters' grip on oil production.

But what about vertical dis-integration — that is, the separation of crude oil production from refining — and the reliance on arms-length transactions to exchange crude internationally?

Luciani has dealt with this question in the past, and shall not burden the reader with a detailed repetition of the argument. Writing in 1984, he argued that the industry would tend to reintegrate, because downstream operators would try and widen their equity base upstream, while major producing countries could not afford to depend entirely on the vagaries of an unstable crude oil market, and would themselves integrate downstream.[1] In a more recent work, Luciani has reviewed developments up to 1990, and considerably revised the argument, taking into account a number of complexities that have arisen in the real world.[2]

Vertical Integration

We shall keep to the definition of vertical integration that was given in the latter article, that is, "the extent to which corporations controlling the output of some of the major oil provinces in the world sell crude as such or prefer to refine it for themselves and sell it as oil products." The location of refineries — whether they are closer to the market or closer to the producing fields — is considered irrelevant.

Even so, available data only allows for very imprecise measurement of vertical integration. It is limited in that it only indicates equity-crude production and the refinery capacity of each company: companies may well be selling their equity crude and buying other crude to refine, using third-party refineries to refine their crude, or making their refineries available to third parties. There is no way of having an exact picture of the degree of vertical integration in the industry, and for reasons that were discussed earlier, we believe the available data underestimates the effective degree of vertical integration. However, there is no way to prove this.

On the basis of these premises, previous work estimated in 1992 an "imbalance index," defined as the ratio of the difference between oil produced and total refinery runs, divided by whichever of these two is larger. With this definition, a positive value indicates a company that produces more crude than it refines, while a negative value indicates a company that refines more crude than it produces. For all positive values, the denominator is production; for all negative values, the denominator is refinery runs. The index thus defined will have a maximum value of 100 for a company that only produces crude, and has no refinery activity; a minimum value of -100 for a company that only refines, and produces no crude; and values close to 0 for companies that are vertically balanced, meaning they produce about as much oil as they refine.

The Top 50

Table 1.1 *Petroleum Intelligence Weekly*'s Top 50 for 1990/1995: Liquids Production and Refinery Capacity

		1990 Liquids Production	1990 Refinery Capacity	1995 Liquids Production	1995 Refinery Capacity
1	Saudi Aramco	6279	2113	8585	2000
2	PDVSA	2135	2298	2885	2402
3	RD Shell	1890	5476	2254	3594
4	NIOC	3183	766	3720	1092
5	Pemex	2974	1679	2722	1520
6	Exxon	1712	4061	1726	3422
7	Mobil	779	2069	810	2256
8	Chevron	935	2154	1001	1665
9	KPC	1042	902	2070	955
10	BP	1322	2053	1213	2000
11	Sonatrach	1063	475	1283	462
12	Pertamina	761	751	1065	986
13	Amoco	782	984	660	988
14	Texaco	810	1448	762	1572
15	ENI	458	717	612	1006

Table 1.1 continued

		1990 Liquids Production	1990 Refinery Capacity	1995 Liquids Production	1995 Refinery Capacity
16	Petrobras	653	1466	716	1540
17	INOC	2125	550	600	603
18	ADNOC	1128	162	1300	211
19	CNPC			2796	496
20	EGPC	571	465	446	548
21	LNOC	1041	534	1345	342
22	Elf	518	560	764	870
23	Total	420	887	456	919
24	Arco	705	534	650	522
25	NNPC	1199	401	1200	400
26	Sidanko			459	793
27	Surgutneftegaz			669	388
28	Conoco	359	579	414	621
29	Yukos			722	657
30	Lukoil			1116	470
31	QGPC	467	70	475	63
32	Phillips	372	305	237	345
33	Rosneft			257	235
34	Marathon	197	638	205	570
35	YPF	293	431	373	325
36	Unocal	254	415	240	251
37	Amerada Hess	175	575	260	575
38	Tyumen Oil			456	361
39	ONGC	656	74	633	
40	Statoil	430	200	499	224
41	Sibneft			409	526
42	Gazprom			18	
43	Petrofina	116	615	141	641
44	Ecopetrol	263	232	460	256
45	Slavneft			258	689
46	Petronas	373	30	370	140
47	PDO	391	71	481	80
48	Petroecuador	203	120	306	142
49	Petro-Canada	94	319	79	285
50	BHP	233	95	208	95

Table 1.1 shows the 1995 list of the 50 most important oil companies, as ranked by *Petroleum Intelligence Weekly (PIW)*.[3] We have assumed this table as the empirical foundation of our analysis. It should be noted that *PIW* makes a concerted effort to provide company-based data (rather than country-based) that includes state companies of the oil-exporting countries alongside major international companies. Yet it should be acknowledged at

the outset that such data is not without question; other sources would show significantly different numbers, in particular with respect to refining capacity. The root of such differences is probably found in terms of definition, for example whether the capacity of a refinery is judged by its maximum theoretical atmospheric distillation capacity, or by effective maximum "balanced" capacity of the plant; the two may differ quite significantly. Another important difference is whether the capacity of a refinery is attributed to a company only if the latter has a controlling interest in the plant, or, alternatively, minority stakes are translated into pro-quota attribution of capacity to the various owners. A significant example is given by the *Oil and Gas Journal*'s survey of the world's top 20 refining companies,[4] which reveals relevant differences in distillation capacities compared to the *PIW*'s survey for the same year, as shown in Table 1.1a. In short, *PIW* data is questionable; nevertheless, it is still the best available for our purposes.

Table 1.1a *Oil and Gas Journal* 1995 Refining Survey: How the World's Largest Refiners Rank

Refining capacities ('000 barrels/day)		*O&GJ* Survey Capacity as of Dec. 31 1995	*PIW*'s Survey	Differences *O&GJ* vs *PIW*
1	RD Shell	4460	3594	866
2	Exxon	3415	3422	-7
3	Sinopec	2867		
4	PDVSA	2500	2402	98
5	Mobil	2044	2256	-212
6	BP	1812	2000	-188
7	Pemex	1631	1520	111
8	Saudi Aramco	1521	2000	-479
9	Chevron	1497	1665	-168
10	Petrobras	1233	1540	-307
11	Texaco	1168	1572	-404
12	NIOC	1167	1092	75
13	Amoco	995	988	7
14	ENI	984	1006	-22
15	Kuwait NPC	934	955	-21
16	Total	921	919	2
17	Pertamina	805	986	-181
18	Idemitsu K	785		
19	Conoco	723	621	102
20	Sun	692		

Based on *PIW* data, imbalance indexes were calculated, and are shown in Table 1.2. The right-hand column shows the index for 1990, the year for which our previous analysis was conducted; the left-hand column has results for 1995. However, the values for 1990 are not immediately comparable to results obtained in 1995, because the composition of the list of companies has greatly changed, and this in itself is an important factor in our discussion.

Table 1.2 Imbalance Index for *Petroleum Intelligence Weekly*'s Top 50

	1995 Indicator	1990 Indicator		1995 Indicator	1990 Indicator
ONGC	100.00	88.72	Rosneft	8.56	
Gazprom	100.00		Pertamina	7.42	1.31
QGPC	86.74	85.01	INOC	-0.50	74.12
ADNOC	83.77	85.64	Unocal	-4.38	38.80
PDO	83.37	81.84	Elf	-12.18	-7.50
CNPC	82.26		EGPC	-18.61	18.56
Saudi Aramco	76.70	66.35	Sibneft	-22.24	
LNOC	74.57	48.70	Phillips	-31.30	18.01
NIOC	70.65	75.93	Conoco	-33.33	38.00
NNPC	66.67	66.56	Amoco	-33.87	-20.53
Sonatrach	63.99	55.32	RD Shell	-37.28	-65.43
Petronas	62.16	91.96	ENI	-39.17	-36.12
Lukoil	57.89		BP	-39.35	-35.61
Statoil	55.11	53.49	Chevron	-39.88	-56.59
BHP	54.33	59.23	Sidanko	-42.12	
KPC	53.86	13.44	Exxon	-49.56	-57.84
Petroecuador	53.59	40.89	Total	-50.38	-52.65
Ecopetrol	44.35	11.79	Texaco	-51.53	-44.06
Pemex	44.16	43.54	Petrobras	-53.51	-55.46
Surgutneftegaz	42.00		Amerada Hess	-54.78	-69.57
Tyumen Oil	20.83		Slavneft	-62.55	
Arco	19.69	24.26	Marathon	-64.04	-69.12
PDVSA	16.74	-7.09	Mobil	-64.10	-62.35
YPF	12.87	-32.02	Petro-Canada	-72.28	-70.53
Yukos	9.00		Petrofina	-78.00	-81.14

PIW's list for 1990 did not include any of the oil companies from the former Soviet Union or the People's Republic of China. For 1995, we found ten companies that remained unlisted, namely: Gazprom, Lukoil, Surgutneftegaz, Tyumen Oil, Yukos, Sibneft, Sidanko, Rosneft, Slavneft, and the China National Petroleum Corporation (CNPC). Ten companies have left the list: Sun, Occidental, Oryx, Banoco, British Gas, Ultramar, Coastal, Norsk Hydro, Repsol, and Wintershall. Accordingly, the significance of the *PIW* 50 group of companies has increased: the 50 largest companies of 1990 accounted for approximately 40 million barrels per day (mb/d) of liquids production, while today's 50 largest account for 51mb/d.

The weighted average (using liquids production as the weights) of the imbalance index in the 1990 group was 53.7, indicating a prevalence of vertically nonintegrated companies. The same average as calculated for the 1995 group of companies is 31.1, indicating that the 1995 group is characterized by a lesser degree of vertical imbalance.

Table 1.2 indicates that the overall vertical balance of the international oil market was subjected to contrasting forces, some increasing, others decreasing, the overall balance. Among factors decreasing the balance, we may note the following.

- The entrance in the *PIW* 50 list of a group of severely unbalanced companies, such as CNPC and most of the Russian companies. Their imbalance has to do with their formation process — in the case of the Russian companies — and with the fact that China's domestic-oil market is still centrally planned and insulated from market forces. We expect the Russian oil companies to evolve rather rapidly towards greater vertical balance through the exchange of assets and various international joint ventures, but in fact, the implementation of market principles in the Russian energy sector has been much slower than was initially expected, and this in turn has hindered the progress towards greater vertical balance.
- The growing liquids production of a small group of producing countries' national oil companies (NOCs), whose downstream integration has not proceeded at the same pace. Companies registering the most important increases include Saudi Aramco (2.306mb/d), Kuwait Petroleum Company (KPC) (1.028mb/d), Petróleos de Venezuela SA (PDVSA) (750 thousand b/d), National Iranian Oil Company (NIOC) (537tb/d), Libyan National Oil Company (LNOC) (304tb/d), Pertamina (304tb/d), Sonatrach (220tb/d), Ecopetrol (197tb/d) and Abu Dhabi National Oil Company (ADNOC)

(172tb/d). Of these, Saudi Aramco, Pemex, and LNOC registered a decline in their overall refining capacity, while the more significant increases were those of NIOC, PDVSA, and Petronas.

It should, however, be noted that the NOCs of some producing countries have invested downstream, but simply did so at a slower pace than their respective production's growth. Table 1.3 lists recent deals involving the sale of refining assets; indeed, a number of NOCs are listed as buyers in these transactions. The reader should not remain under the impression that nothing is happening in the market, as this is certainly not the case.

Table 1.3 A Profile of the Main Upstream and Downstream Deals in 1995–96

	Buyer	Seller	Assets	US$m
1995	YPF	Maxus Energy	All properties in Latin America and Indonesia	1,735
1995	Statoil	Aran Energy	All properties	286
1995	ENI/Conoco/Shell	Czech Gov.	49% in two refineries, Litvinov and Kralupy	175
1995	S. Aramco	Vardinoyannis Gp	50% of Motor Oil Hellas and 650 gasoline stations (Corinth refinery and Avinoil)	430
1995	Australian Petrol	Pioneer/Caltex Australia	Merger with a US$115m adjustment for Pioneer (Lytton and Kurnell refineries)	
1995	Finpetro	Total	48% of Petrocontrol, which holds 25% of Petrogal (Porto and Sines refineries)	na
1995	Nippon Oil	Caltex	50% stake in Nippon Petroleum Refining	1,270
1996	Statoil	Conoco	Takeover of JET's 257 filling-stations in Ireland	na
1996	ÖMV	KPC	Q8 service chain in Hungary	na
1996	Repsol	Astra	37.7% oil and gas company in Argentina	360
1996	Petronas	Engen	30% stake in the South African refiner and marketer	436
1996	PDVSA (Lagoven)	Mobil	50% ownership of Chalmette Refinery in Louisiana, and development and upgrading of extra-heavy crudes of Orinoco Oil Belt (The scheme requires Congressional approval)	na
1996	KPC	ENI	50% stake of Milazzo refinery in Sicily and 336 filling stations in Italy	na

Increasing Vertical Integration — Who and What's Involved?

Among the factors increasing vertical balance, one should include the behavior of several major international oil companies. On the one hand, they have significantly increased their crude-equity production, due mainly to successful exploration and improved recovery technology; on the other, they have shown a clear tendency to reduce their presence in refining. The first and foremost example of this strategy is Shell, which has increased its equity production by 364tb/d, and reduced its refining capacity by an astonishing 1.8mb/d. Consequently, Shell's imbalance index almost halved, passing from -65 to -37. Other major cuts in refining capacity were made by Chevron and Exxon, each in the range of 500tb/d, accompanied by a small increase in crude production. This trend continued through 1996, further increasing vertical integration.

The major companies' equity production will also increase as a result of developing a closer working relationship with several of the major oil-producing countries. The cases of Algeria and Venezuela are particularly notable, as companies have made significant discoveries there, and in a few years, equity flowing from those countries will almost certainly play a significant role in their total production. The process may be reinforced by Kuwait and — if and when sanctions are fully lifted — Iraq. Iran is also experimenting with a timid form of opening. Saudi Arabia stands out as the major exception.

Since 1987, Saudi Aramco has had a major downstream joint venture with Texaco in the US, Star Enterprise. More recently, it has acquired interests in three refining companies, one each in South Korea, the Philippines, and Greece. Hence, Saudi Aramco's international refining presence has increased slowly. However, its domestic refining expansion plan has been suspended, and the capacity of the Ras Tanura refinery decreased. Saudi Aramco's international presence in refining will further increase in the coming months in connection with some of the ventures in which it is involved, specifically the US$17 billion venture that will combine the domestic refining and marketing operations of Shell and Texaco, creating the largest marketer of oil products in the US. The joint venture will be 35 percent owned by Shell, 32.5 percent by Texaco and 32.5 percent by Saudi Refining, an affiliate of Saudi Aramco. Such a merger will undoubtedly lead to an expansion of the refining capacity attributable to Saudi Aramco. The company is also known to be negotiating the acquisition of a 35 percent stake in Petrogal-Petroleos de Portugal. It has also expressed an interest in more distant and speculative projects, such as refining in China.

PDVSA is possibly the one state company that has most consistently pursued a policy of downstream integration. Its strategy has been to increase market share for refined products while maintaining crude supplies to traditional customers. This policy has resulted in expanding domestic refining capacity as well as holding interests in approximately 15 refineries outside Venezuela, mainly in the US, Germany, Sweden, and Britain. In October 1996, a further agreement with Mobil to acquire a 50 percent interest in the 176tb/d Chalmette refinery was announced. The ultimate goal is to ensure stable demand for the development of its extra-heavy crudes from the Orinoco Belt, and to profit from the margins of higher-value-added refined products.

Another company significantly geared towards downstream integration is KPC, which made its first investment in Europe in 1983. Today, its network comprises approximately 6,500 service stations, with sizable market shares in Denmark, Sweden, Italy, Belgium, Britain, and the Netherlands. The company owns two refineries, in Holland and Denmark (though the latter is being used for storage only). In Italy, after the closure of its Naples refinery in 1993, KPC finalized a deal with ENI for a 50 percent stake in the Italian Milazzo refinery (capacity 160tb/d) and the parallel acquisition of 336 filling stations. In Hungary, KPC has not had the opportunity to integrate its retail network with the acquisition of a local refining interest, as the privatization process has made little progress so far. Consequently, it decided to leave the market, selling its assets to Austria's Österreichische Mineraloel Verwertung AG (ÖMV). In Asia and the Far East, where KPC started operating only in the 1990s (mainly Thailand and Singapore), the company hopes for further expansion; negotiations are underway for the possible development of new refining joint ventures in Pakistan, India, Malaysia, and Indonesia.

ADNOC distillation capacity will increase from 210 to 480tb/d, with the addition of the two planned condensate units at Ruwais. This very ambitious expansion project leaves less interest for international expansion in refining. ADNOC is nevertheless said to be interested in a refinery project in Pakistan. Abu Dhabi's International Petroleum Investment Company (IPIC) has minority stakes in the Spanish company, Cepsa, and in ÖMV. As IPIC is different from ADNOC, this downstream investment cannot be attributed to the latter.

Libya started establishing its presence in the European downstream market in the late-1980s, mainly in Italy, Switzerland, Germany, and Holland, with minor distribution interests in Spain, Hungary, the Czech Republic, and Slovakia. However, to protect investments from UN sanctions, Libya relinquished control of the holding company in 1993, retaining a 45 percent interest and selling the majority stake to a group of Italian and German private investors. In any case, this investment was not directly linked to the LNOC, and is therefore not included in this analysis.

NIOC's domestic refining capacity grew in 1993–94 with the start-up of the Arak refinery, but domestic self-sufficiency in petroleum products will only be reached with the complete start-up of operations of the Bandar Abbas refinery. In 1970, NIOC acquired a 17.5 percent interest in South Africa's Sasoburg refinery, but in 1979 relations between the two countries were severed. Since the re-establishment of diplomatic and economic links in 1994, negotiations have started for new-term crude oil sales.

Petronas has recently intensified its push towards international expansion, both upstream and downstream. In order to increase its retail and refining activities abroad, Petronas acquired a significant stake (30 percent) of Engen, the largest South African oil marketer. The company operates an 85tb/d refinery in Durban, and markets products in South Africa, Namibia, Botswana, Lesotho, and Swaziland. It is certainly poised to become a major player in the game of downstream integration and progressive transformation of national state oil companies into integrated international majors.

A similar example in Europe is Statoil, which continues to expand its European downstream operations, with the intention of widening its export markets. After the acquisition of British Petroleum's Swedish network in 1993, becoming the leading retailer in Scandinavia, Statoil has recently focused its attention on Ireland, with the major takeover of Aran upstream activities and the purchasing of Conoco's Irish retail network ('Jet' brand).

Pemex has been slow to develop a strategy of foreign ownership for refining and marketing assets, and today its only equity stakes are in the US (Shell) and in Spain (Repsol).

Sonatrach has not increased its refining capacity since 1980–81, when the large Skikda refinery and the small plant at In Amenas came respectively onstream.

Loose Oil

On the basis of *PIW*'s data, an estimate of what we refer to as "loose oil" is also possible. Loose oil is the difference between crude oil production and refining capacity, indicating the amount of crude that is necessarily exchanged at arm's length. This concept obviously has meaning only for those companies for which crude production exceeds refining capacity. Results are shown in Table 1.4 and Fig. 1.1. In the 1995 group of companies, total loose oil amounted to 21.84mb/d, while in 1991 and for a different group, we found a total of 19mb/d. The limited increase is well justified by the inclusion of companies that are almost exclusively upstream.

Table 1.4 "Loose Oil" for *Petroleum Intelligence Weekly*'s Top 50 (1995)

	tb/d		tb/d
Saudi Aramco	6,585	Surgutneftegaz	281
NIOC	2,628	Statoil	275
CNPC	2,300	Petronas	230
Pemex	1,202	Ecopetrol	204
KPC	1,115	Petroecuador	164
ADNOC	1,089	Arco	128
LNOC	1,003	BHP	113
Sonatrach	821	Tyumen Oil	95
NNPC	800	Pertamina	79
Lukoil	646	Yukos	65
ONGC	633	YPF	48
PDVSA	483	Rosneft	22
QGPC	412	Gazprom	18
PDO	401		
		Total	**21,840**

Source: *Petroleum Intelligence Weekly*

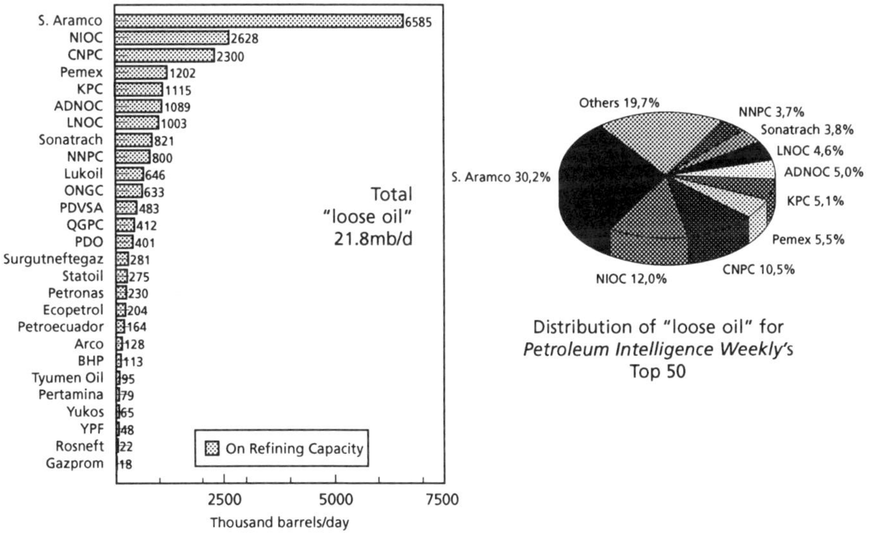

Source: *Petroleum Intelligence Weekly*

Fig. 1.1 "Loose Oil" for *Petroleum Intelligence Weekly*'s Top 50 (1995)

As Fig. 1.1 clearly indicates, the majority of loose oil is in the hands of Saudi Aramco, and much of the future of the entire process of vertical reintegration depends on whether or not it chooses to pursue downstream integration with greater vigor. Also note the importance of NIOC, whose downstream integration is presumably made much more difficult by the US-imposed embargo. Finally, note the importance of CNPC, whose oil is in fact not loose at all, as it is fully committed to the Chinese domestic market, and exchanged with state-owned refineries on the basis of an administrative rather than market-based process.

There is no reason to expect vertical reintegration to be pushed to such an extreme that it entirely eliminates loose oil. Indeed, it is almost certain that this will never happen. The group of major producers that are still severely out of balance is limited; in fact, just two companies stand out as the main contributors to the overall statistical result.

The Pace of Integration

Why is it that vertical integration has not progressed further? Why do some NOCs of the producing countries hesitate to engage in downstream investment?

There are a variety of simple and complex answers to this question. Among the former is the fact that in a period of low oil prices and significant budget deficits, governments that own NOCs have a high priority for a larger cash flow, to the relative detriment of the companies' investment budgets. Also, these same budgets have had to face growing requests for maintaining or increasing production capacity, meaning that NOCs felt they could not afford a more aggressive downstream investment policy.

A Downstream Presence

Is downstream investment an attractive proposition? If NOCs were really convinced of the need to integrate downstream, budget constraints would have been loosened in order to allow for a more aggressive integration policy.

The fact that some of the major NOCs have been reluctant to accelerate their presence downstream is an indication of their lack of enthusiasm for the prospect. If one takes a short-term view — and assumes that crude and products prices are given, and independent of the degree of vertical integration in the industry — then skepticism *vis-à-vis* downstream investment is well grounded.

Indeed, the return on capital invested in developing new oil fields or enhancing recovery in old fields is a multiple of anything that can be obtained through investment in refining. In fact, most companies have been steadily reducing their presence in refining, and have attempted to cut costs through mergers and alliances. The process is slowed down by the environmental costs involved in closing down a refinery, and by the concern that a refinery sold to an independent operator at a knocked-down price may turn into a dangerous competitor, making matters even worse.

Refining margins also vary according to regional factors; Fig. 1.2 shows refining margins in Europe, the US, and the Far East, and clearly illustrates that European margins are low, American margins better, and Asian margins very attractive.[5] Yet in terms of volume, the most important markets for Gulf producers are the OECD countries of Europe and Japan, i.e. precisely those countries in which, notwithstanding several attempts, inroads have proven most difficult.

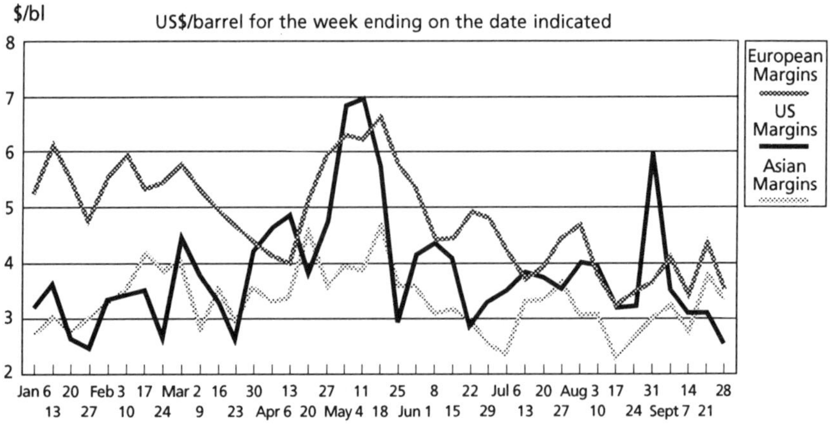

Source: Salomon Brothers' weekly *Domestic Regional Analysis of Margins* (DREAM) and *Margin Analysis of Global International Countries* (MAGIC)

Fig. 1.2 Weekly US, European, and Asian Refining Margins (January-September 1996) (value of refinery products at the refinery-gate minus the price of crude)

The high-margin markets in South-East Asia are small. Gulf NOCs are investing there, though such activity will make little difference to the oil industry's overall level of vertical integration in the short run: it is investment aiming at future growth. Furthermore, the two largest and fastest-growing markets in the region, i.e. China and India, have not been sufficiently forthcoming in attracting foreign investors, and most of the projects under consideration are still at the planning stage.

NOCs for which the US is an important market — most notably PDVSA and Saudi Aramco — have invested in establishing a downstream presence there. For other companies, the American market is not sufficiently important — at least for the time being — to justify investing in US refining.

This is why, in essence, the issue boils down to insufficient downstream integration in Europe and Japan. For the latter, the obstacle may not be purely economic, but for Europe undoubtedly, low refining margins have discouraged investment. It should be noted that investment has taken place nevertheless, but not at the pace required to achieve more significant vertical reintegration.

The State of Refining

1995 was the worst year for refiners since 1987, with many plants losing significant amounts of money. Return on fixed assets in global downstream activities was estimated at only 6 percent for the year, compared to 12–14 percent in 1990–91. Only the relative health of marketing earnings has kept overall downstream returns out of the red.

Although a structural turn in the cycle is not expected very soon, the sector in perspective does offer great potential for improvement and profitability. Demand continues to rise, and with the closure of loss-making refineries, capacity utilization is increasing almost everywhere. The world's current consumption of crude oil is nearing 71mb/d, and the present world's refining capacity is 74mb/d. There are approximately 700 refineries around the world, but the out-turn varies from region to region.

In Europe (113 refineries, 14.2mb/d), declines in fuel-oil demand and new gasoline specifications led to a surge in investment in upgrading capacity. Much of this new capacity went onstream in recent years, and was hit by recession and a shift toward diesel cars. The result has been a surplus of gasoline, with the situation worsened by the introduction of reformulated gasoline (RFG) in the US (representing a new barrier to entry in the US market, the largest for European gasoline exports) and the reduction of price differentials between light–sweet and heavy–sour crudes, due to another steep increase in North Sea production and a drop in exports of sour crudes.

In the US, where crude production is expected to fall, we estimate a rising product demand of more than 1.5 percent annually during the second half of the 1990s.

In Asia and the Far East (138 refineries, 14.8mb/d), there are few doubts about rising demand and the need for additional capacity. The problem is that nobody wants to miss this opportunity, and so there are fears of possible oversupply. Consequently, the outlook for the region could be less positive than present-day levels.

Refining margins in Europe stay low because of declining gasoline demand, and the difficulty of closing down old and obsolete plants. Refiners are caught in a difficult position between environmental regulations mandating increasingly costly investment, and a market that is not price-responsive to quality, partly because of the extraordinary burden of excise taxes. This creates an ambivalent situation in which, on the one hand, it is not too difficult to acquire refining properties, as several of the present owners would be happy to cut back their presence in the sector; on the other, prospects for future profit are less than clear.

Eventually, excess refining capacity will be closed down, and lower-quality products will be driven out of the market, either by consumer awareness or by administrative intervention. Refineries will then be very different from what they are today — closer to a petrochemical plant than to a simple atmospheric distillation column — and will again be profitable. At that time, the cost of entry — financial and technological — may simply be too high even for the NOCs of the major producing countries.

Downstream Investment and Today's Market

Currently, downstream investment remains a rather unattractive prospect from a purely financial point of view. Why should NOCs aggressively pursue downstream integration?

The rationale has little to do with refining returns and a lot to do with the organization of the international oil market. In other words, we claim that if NOCs were integrated downstream, and selling final products rather than crude oil, the behavior of the oil market would be different, and more favorable to them.

Today's international oil market is a paradox. On the one hand, it is true that a market exists, and oil can be readily purchased by anyone who cares to do so at the going price. This is a major step forward with respect to conditions prevailing at the time of the Seven Sisters. Yet price determination on this market leaves a lot to be desired.

Price is in fact determined by two crude flows, Brent and West Texas Intermediate (WTI), that are very small relative to the huge quantities that are exchanged internationally at arm's length. The vast majority of crude oil transactions take place at prices that are related to Brent or WTI via a formula that is not influenced by supply and demand for the specific crude that is being traded. The international oil market thus has the shape of an inverted pyramid; indeed, a substantial number of transactions occur at prices that are determined by the demand for and supply of two marginal crudes.

The consequences of this state of affairs were analyzed by Mabro and Horsnell in their seminal work.[6] We have little to add to their study, except in pointing to how the situation would evolve through aggressive downstream integration.

The fact that the price of Brent does not reflect so-called fundamentals is by now a well-known paradox. In theory, the final demand is for petroleum products; their availability should influence available stocks and refinery runs. In turn, demand from refiners and their preference for holding crude stocks should guide the crude market. Further, in theory, trading in Brent should reflect marginal demand and marginal supply, justifying this market's role as price-maker, notwithstanding its limited size.

In practice, the price of crude is largely arbitrary, within a wide tolerance band. The Brent market is not marginal at all, in the sense that it is not predominantly affected by shifts in supply and demand for crudes different from Brent. Rather, it is influenced by specific circumstances that have nothing to do with the overall equilibrium of the global oil market. Whatever happens, the price of crude is transmitted — possibly with some delay — to the price of products.

By integrating downstream, NOCs would be able to bypass the crude market and sell directly into the products market. Products markets are more liquid and transparent than crude markets, and better reflect the balance between supply and demand. If a large supply of products became independent of the vagaries of the crude market, a gradual decoupling of the two would take place. Because Brent could not possibly afford to deviate too much from the netback realized price of products, the causation link would be inverted.

We maintain that NOCs would profit by integrating into products markets, because once a company has established its presence downstream in a geographically defined market, the limitation of logistics will protect its trading position. Oil refining and products marketing is a territorial business.

It is well known that price competition at the gas pump is of limited appeal. What this means is that the retail price in most markets does not need to follow the price in Rotterdam immediately, which in turn does not need to reflect Brent mechanically. The news that Iraq may be allowed to export some oil does not need to affect the price of gasoline in Manchester immediately. In other words, the market for crude may become long or short, while the market for products wholesale does not do so to the same extent. The retail market for products is also not fully responsive to changes in the wholesale market. Therefore, there may be times when crude is best sold as crude, and other times when it is best sold as wholesale products (changing refinery margins are an indication of this); the retail market provides operator stability.

NOCs of the major oil-producing countries need to refine their activities in this respect. They could optimize their trading position if they were present in all markets, selling crude on the spot market, selling products wholesale, and through their own retailing networks. They could, in this way, enjoy volume stability to the extent that they may occupy a share in the retail market, reap the benefit of refining margins whenever these are positive, and modulate their sales on the crude market in order to resist short-term price changes that may not be linked to fundamentals.

An international market that was more responsive to fundamentals would work better than at present. Yet this objective is unlikely to be reached unless major crude producers integrate downstream, because demand for crude is remote from the final consumers and subject to the pressure of speculation, in a context in which, furthermore, the major producers have chosen to be price-takers rather than price-makers.

The argument is further reinforced by a consideration of the changing nature of the refining industry. As mentioned before, the demand for products with superior environmental qualities is rapidly increasing. Gasoline and other fuels are less and less of a petroleum product and more and more of a petrochemical one. Examples include the elimination of lead, reducing and minimizing aromatics and oxygenates, and the need to maintain high performance as measured by the octane number. In short, present-day fuels are complex mixes of different components, some of which may not be oil products at all.

The industry in Europe is engaged in developing a large "Auto-Oil Program" that will develop better fuels and better engines, further underlying the importance of quality. Environmental concerns are here to stay, and will probably grow: the only alternative to constantly improving quality is cutting back on utilization. The latter is certainly not in the best interests of major oil producers.

What Does the Future Hold?

There are two ways in which this long-term trend affects our argument. The first is that the price of petroleum products will grow increasingly decoupled from the price of crude. Crude will be just one of the raw materials utilized in a sophisticated industry, whose final pricing strategy will depend as much on financial considerations (the cost of capital is key in highly capital-intensive ventures), the cost of other inputs, and marketing strategies.

Hence, to be a pure producer of crude will become less and less interesting. Notwithstanding the fact that today's refining margins are low, in the future, a growing share of the pie will undoubtedly be captured by industrial transformation downstream.

Buying into this process — acquiring familiarity with the technology, establishing control of a market share — will be increasingly expensive in the future. As the industry is transformed from production of commodities through generally available technology, into production of branded quality products through sophisticated and proprietary technology, entry barriers will be increased. Major oil producers should be aware of the process and actively participate from the beginning if they wish to make sure that they will continue to understand fully what the petroleum industry is about.

The simple advantage of controlling major oil resources is becoming less and less important. Oil reserves are being extended even in older provinces, and the danger that the time of exploiting some of the more abundant reserves in the Gulf may be indefinitely postponed is real. This danger is only avoided through integration downstream and access to the consumer.

The industry has witnessed quite radical changes in the last ten years, and nothing indicates that the rapid pace of change is slowing. Some of the changes will enhance vertical integration, others will go in the opposite direction, but overall it is easier to expect events in the former category than in the latter. Hence, we expect that the vertical balance of the international oil market will progressively improve, and that major producing countries will increasingly come to the conclusion that integrating downstream is not a matter of convenience, but an obligation.

CHAPTER 2

International Downstream Integration of National Oil Companies

Majid A. Al-Moneef

Introduction

The oil market has been characterized by various dichotomies that have affected not only its stability but also the relationships among its participants. On the institutional side, the OPEC/non-OPEC — or producer /consumer — have shaped oil-market relations over the past two decades. On the functional side, the fundamentals versus technical relations still affect the analysis and understanding of the oil market. On the industrial front, the distinction between the majors and the independents, and between national oil companies (NOCs), i.e. those owned by oil-producing states and their international counterparts, continues to affect the structure of the industry and its investment options.

Many of these dichotomies are illustrated in the relationship between the upstream and downstream ends of the industry. Such relations have had a profound effect on the industry, the market and ultimately on producer-consumer ties. Although the 1970s witnessed the growth of OPEC, particularly in its role as arbiter of production quotas and pricing decisions, it was also characterized by the continuing disintegration of the international oil industry, a process that had started two decades earlier by the entry of the independents.

It should be pointed out, however, that vertical integration has come to characterize the growth and dominance of the industry. Indeed, vertical integration has directly contributed to the growth in oil demand. It should also be noted that the independents' aggressive entrance into the market in the 1950s and 1960s, and the producer-governments much-publicized control in the 1970s, did not result from a vertically integrated market per se, but rather from the oligopolistic nature of the industry on the one hand and the old concession arrangements on the other. The diminishing role of the international oil majors in the 1970s led to an abrupt separation between upstream and downstream activities, an aspect that ultimately affected the stability of the market, and the relations among its participants. Indeed, the developments of the forward and futures markets can be traced back to such segregation.[1]

Other factors that contributed to the disintegration of the industry included: government controls, environmental legislation, the development of new producing areas, and the over-politicization of oil relations throughout the 1970s and early 1980s. Such disintegration posed several challenges and opportunities to the oil industry. Many international oil companies underwent a process of restructuring that affected business lines and geographic activities. Some added new equity oil to their operations, whether through developments of new producing areas to substitute for the lost equity oil from the Middle East and Venezuela, or through acquisitions. Others were not so fortunate, and in order to adapt to the new market realities, many were forced to scale down their operations and emphasize one line and/or geographic area.

The process of disintegration throughout the 1970s also witnessed the emergence of NOCs. Companies such as Kuwait Petroleum Company (KPC), Petróleos de Venezuela SA (PDVSA), Petróleos Mexicanos (Pemex), Saudi Aramco, and Statoil became active participants in the market. This group of companies assumed control of large oil and gas reserves, and had either total or shared control over the production and marketing of crude oil. There were, of course, differences in the business and market experience of each of these NOCs; they did, however, share a number of common features such as a quasi-domestic monopoly over the extraction of oil and gas resources, full governmental support, and the powerful role they performed in their respective economies. Once this group became established in the international business arena — and given the changes in the market — they too were confronted by exactly those challenges and choices faced by other large oil companies. Indeed, the move away from administered prices, as well as the growing competitive aspects of the market, affected the strategies and alliances of NOCs.

The Reintegration of the Oil Industry

The integrated structure of the oil industry experienced considerable strain in the 1970s and early 1980s. The loss of control over oil pricing and production decisions by the major oil companies, combined with nationalization and take-overs, effectively deverticalized the industry. Reserves and a large share of the upstream phase of the industry came under the control of NOCs, while refining and product marketing in the main consumer areas remained largely with the international oil companies. Since the governments of the NOCs were not directly involved in the market for final products, they had little knowledge of the impact of their upstream pricing decisions on that market.

There were other forces at play in the oil market that affected the industry and its choices. For example, oil substitution, and consumer government policies reduced oil's share in the global energy market, while supplies from alternative sources contributed to OPEC's declining share in total world oil consumption. Such trends ultimately affected the power and cohesion of the OPEC cartel.[2]

On the industrial structure side, major changes in the upstream and downstream patterns were under way. Seizing on the opportunities created by higher oil prices, international oil companies attempted to reposition themselves in the upstream by developing new areas and engaging in reserve acquisitions. The downstream side of the industry was also restructured, as refining and distribution processes witnessed changes in ownership patterns. Throughout the 1980s, companies were consolidated, refining capacities mothballed, and unprofitable business lines shut down.[3]

The NOCs of OPEC did not escape these structural changes; as established major upstream companies, they saw their shares in the world crude market dwindle as more expensive reserves replaced theirs in major markets. Such a reduction was partly due to higher prices resulting from the prevailing market structure, as well as the disintegrated market structure. During the 1980s, some NOCs joined the process of restructuring, and embarked on increasing their downstream presence. Having built adequate domestic refining capacities, they seized the opportunities presented by the downstream restructuring, which was underway in Europe and the US, and purchased refining and distribution facilities in these markets. Thus, a new process of verticalization of the industry began, with NOCs as major players.

The first acquisition of refining and distribution facilities by an oil producer started in 1983, when KPC purchased, through its subsidiary Kuwait Petrochemicals Industries (KPI), a portion of Gulf Oil's European downstream assets.[4] Other NOCs have followed the reintegration trend and entered into joint ventures, acquired shares, or purchased refining and distribution assets in the major markets of Europe, the Far East, and the US. Table 2.1 highlights the major international downstream ventures of NOCs since 1983.

Patterns of Downstream Integration

Investments in refining and marketing assets in the major markets by oil producers began in the early 1980s, when they were enjoying large revenues from higher oil prices. It also coincided with attractive opportunities which arose during the restructuring phase that international oil companies experienced after the breakdown of the old government-company resource-property relationship, and the drive by such companies to divest themselves of non-profitable business lines. The reintegration drive, however, continued after the oil price collapse and the resulting decline in oil producers' revenues. It is interesting to note that verticalization by NOCs coincided with virtual deverticalization by the international majors, whereby non-profitable assets in the downstream were sold off due to declining refinery margins and capacity overhang in the refining sector.

Table 2.1 Producing Countries' Refining and Marketing Interests Overseas

	Sites	Installed Capacity (mb/d)	Share (%)	Distri-bution outlets	Crude Supply	Companies[4]
Abu Dhabi[1]						
Germany (1990)	Burghausen	72	19.6			ÖMV
Austria (1994)	Schwechat	210	19.6			ÖMV
Spain (1988)	San Roque	205	10			Cepsa
	Tenerife	89	10			Cepsa
	Huelva	100	10			Cepsa
	Asisa	10	10			Cepsa
Total		**686**				
Kuwait (KPC)[2]						
Denmark (1983)	Skaelskoer	59	100		59	KPC
Netherlands (1983)	Rotterdam	80	100	5,400[3]	100	KPC
Italy (1996)[4]	Milazzo	240	50			AGIP
Total		**379**			**159**	
Libya[1]						
Italy (1986)	Cremona	95	45			Tamoil
Germany (1993)	Harburg	78	30	2,572	150	Holborn
Switzerland (1990)	Collombey	72	45			
Greece		90	14			
Total		**335**			**150**	
Saudi Arabia (Aramco)[2]						
USA (1988)	Delaware City	150	50			Star Enterprise
	Convent	220	50	10,000	467	Star Enterprise
	Port Arthur	245	50			Star Enterprise
Total		**615**			**467**	
S. Korea (1991)	Onsan	525	35	1,125	472	Ssangyong
Philippines (1994)	Bataan	165	40	964	123	Petron
Greece (1995)	Corinth	100	50	619	na	M Hellas
Total		**790**			**595**	
Venezuela (PDVSA)[1]						
USA (1983-87)	Lake Charles	320	100	13,000	170	Citgo[5]
	Corpus Christi	165	100		140	Citgo
	Lemont	150	50		135	Uno-Ven
	Seaview	84	100		30	Citgo
	Houston	265	30		130	Lyondell
	Savannah	28	100		28	Ex-Amoco
Total		**1,012**			**633**	

Table 2.1 continued

	Sites	Installed Capacity (mb/d)	Share (%)	Distribution outlets	Crude Supply	Companies[4]
Germany	Sholven	110	50			Ruhr Oel
	Horst	100	50		145	Ruhr Oel
	Neustadt	144	12.5			E.R.N.
	Karisruhe	142	16.5			ÖMV
	Schwedt	220	18.7			Schwedt
Total		716				
Belgium	Antwerp	15	50			Nynas
Sweden	Nynashamm	25	50		30	Nynas
	Gothenburg	12	50			
United Kingdom	Dundee	10	50			Nynas
	Eastham	22	25		30	Eastham

Notes
1 As of May 1994
2 Figures for 1995
3 Marketing outlets in nine other European countries
4 Concluded in October 1996
5 Citgo is a wholly owned subsidiary of PDVSA. Uno-Ven: Unocal/PDVSA 50/50. Nynas: Neste/PDVSA 50/50. Ruhr Oel: Veba/PDVSA 50/50. ERN: Mobil 50%, Veba 25% and Ruhr Oel 25%. OMW: Ruhr Oel 33%. Schwedt: DEA 37.5%, Ruhr Oel 37.5% and AGIP, Elf and Total 25%. Eastham; AB Nynas 50%, Shell 50%. Star Enterprise is 50% owned by Texaco.

Sources: *Petrostrategies*, May 23, 1994
MEES, various issues

During the 1983–95 period, several downstream deals were concluded between major oil-producing nations, mostly through their NOCs on the one hand, and international companies on the other. This generally involved the purchasing of all or part of the refining and/or marketing assets of the latter, and covering one or more geographic location. As Table 2.1 shows, the scope of the downstream integration varies among countries and across deals.

The minor downstream asset-holdings of Abu Dhabi — for example in Austria, Germany, and Spain — are mostly portfolio investments held by Abu Dhabi's International Petroleum Investment Company (IPIC) and not the Abu Dhabi National Oil Company (ADNOC), although relations between the two have been close, with the latter represented on the IPIC Board. The same situation exists in Libya, whereby investments are held by Oilinvest, and not the Libyan National Oil Company (LNOC). In contrast, all other downstream investments are made through the respective NOCs, with the specific aim of international vertical integration. So is the case for Libya's downstream investments, which are held by Oilinvest and not by LNOC.[5] Except for KPC's sole ownership of the refining and marketing assets of Gulf Oil in Europe, and Venezuela's total acquisition of Citgo's assets in the US, all other deals involve up to 50 percent partnerships with other companies.

Another feature of downstream investments is that for each producing country they involve the traditional market for their crude, and in some cases (e.g. KPC's purchase of Gulf assets in Europe and Saudi Aramco's partnership with Texaco in the US), involve deals with former concessionaires that traditionally processed the selling country's crude. Except for Kuwait's purchase of sole marketing outlets in Italy, the UK, Spain, and other European countries to operate under its own Q8 logo, all other investments involve either refining or refining/marketing assets.

As to the crude-supply arrangements, they differ depending on the type of deal, the configuration of the refineries, and the share of the producing country in the joint ventures. Since crude pricing in virtually all the deals is spot-related, crude supplies from the partner NOC can differ depending on prevailing market conditions. *Petrostrategies* estimated that the crude supply from NOCs to some 25 refineries involved in producer international downstream ventures to be 2 million barrels per day (mb/d) from an estimated name-plate capacity of 3.11mb/d.[6] In the case of Venezuela, for example, crude supply from PDVSA to its totally owned subsidiaries or joint ventures is estimated to be 46 percent of capacity: 838 thousand barrels per day (tb/d) from a possible 1.8mb/d. For Saudi Aramco, the crude supply to its US, Korean, and Philippine refineries in 1995 is estimated to have been 81 percent of overall capacity: 1,062mb/d from a possible 1,305mb/d.

Another feature of overseas downstream ventures carried out by NOCs is the independence of the business line. This is an important characteristic of the investments, not simply from a pure business perspective, but also for the benefits of consumers and producers. From the consumers' side, independence eases the old fear of dominance of the oil producers in the final-products market. As Table 2.2 shows, by the end of 1994, NOCs controlled 9 percent of the overall refining capacities in the US, Western Europe and the Far East (3.84mb/d distributed between 1.64mb/d in the US, 1.5mb/d in Western Europe, and 0.7mb/d in the Far East). The equity share of the oil producers in these refineries is estimated to be 1.84mb/d (1.1mb/d in the US, 0.49mb/d in Western Europe and 0.25mb/d in the Far East), which constitutes only 4 percent of the combined capacities in the three regions. Despite such figures, the concern that oil producers dominate the downstream sector in the consumers' market continues to prevail. The row over KPC's purchase of 22 percent of British Petroleum's (BP) shares in 1989, and the subsequent order by the British government to reduce the holding to 9.9 percent, is an example of such fear. Japan's reluctance to ease restrictions on downstream investments in its market is a further example.

From the producer's perspective, independence of the overseas downstream ventures is important for their global crude oil marketing strategy, and for the efficient operation of the venture as a separate profit center. If the downstream investment is in a joint-venture firm with an established refiner/marketer with knowledge of the market, then independence ensures efficiency and future growth.

Table 2.2 shows the distribution of refining and marketing investments by the five oil producers in the three markets to be 54.7 percent in the US, 32.2 percent in Western Europe, and 12.9 percent in the Far East. Until 1995, the most diversified producers in terms of geographic locations were Saudi Arabia and Venezuela, with 51 percent of Saudi Aramco's equity refining overseas in the US, 41 percent in the Far East, and 8 percent in Europe (Greece). PDVSA has 77 percent of its equity refining capacity in the US, and 23 percent in Europe. All the other producers' equity refining capacities until 1995 were located in Western Europe (Kuwait 25 percent, Libya 16 percent, and Abu Dhabi 15 percent).

Table 2.2 Refining Capacities and the Share of Oil Producers in Major Markets (mb/d)

	US		W Europe		Far East		Total	
	Cap.	Share	Cap.	Share	Cap.	Share	Cap.	Share
Abu Dhabi[2] (205)			686	94			686	94
Kuwait[2] (820)			379	159			379	159
Libya[1] (342)			245	98			245	98
Saudi Arabia[2] (1,670)	615	308	100	50	690	250	1,405	608
Venezuela[1] (1,181)	1,012	751	810	223			1,822	974
Total (4,218)	1,627	1,059	2,220	624	690	250	4,537	1,933
Ref. Capacity	15,129		14,758		14,198		44,085	

Notes
1 Figures for 1994
2 Figures for 1995 (the bracketed numbers indicate domestic refining capacities)

Source: Table 1, OPEC *Annual Statistical Bulletin* (1994)

The Merits and Demerits of Downstream Integration

The advantages of downstream investments by oil producers in overseas markets is recognized by industry representatives and well documented in the current literature. Benefits such as injecting cash into a refining industry in dire need of restructuring, fostering competition in the final-product market, and enhancing supply security to the consumer are said to be the result of such investments.[7] Although vertical integration is recognized as the linchpin of the industry, the scope, extent, and consequences on the market of overseas downstream integration by NOCs is still being debated.

From the industry's perspective, vertical integration, besides reducing costs, seeks to spread risk and capture potential profits at every stage of the chain between well head and gasoline pump. Profits of single phases of vertical integration — namely exploration, production, trading, transport, refining, distribution and marketing — fluctuate asymmetrically. Integration helps to balance a company's operations and protect it from the inherent instability of the market, so when crude prices are low, refining and marketing margins can generally be expected to be positive.

A recent study found that returns from refining and marketing for the period 1980–1990 negatively correlated with returns from the production of crude in six major industrialized countries. The results showed that refining margins increased by approximately 64 cents per barrel for every one-dollar-per-barrel decline in oil prices in Japan, and between 20 and 25 cents in other countries. For a crude price of US$20/barrel (bl), the results estimate a margin of US$30.76/bl in Japan, US$8.32/bl in the US, and US$9.57-9.74/bl in the UK, France, Italy, and Germany.[8]

For oil producers, vertical integration helps capture the value added from refining and marketing the barrel of oil. This value added route was taken by all the producers when they embarked on building refining capacities for local consumption or exports in their own countries. Vertical integration overseas extends this search for global value added, and diversifies the sources of income. It also widens the producers' interests to include the value of the barrel of products sold to the consumer and its determinants, instead of viewing the concentration and definition of interests solely in terms of crude oil price.

A further benefit of vertical integration for oil producers is the potential to secure market share and ensure its future growth. This is important at times when the market is more competitive and refiners seek the most secure source of supply: vertical integration internationally provides this security. This aspect is also important for the exporters of heavy crudes, such as Venezuela and Kuwait, which through investment in high conversion refineries abroad, guarantee an outlet for their crudes.

Another feature of vertical integration that is of interest to oil producers is that the majority of the joint ventures fall into what John Treat and others have termed the industrial "strategic alliance" framework. Such industrial alliances have been increasing at an annual rate of 25 percent. Business alliances such as Disney-Capital Cities, Time Warner-Turner Broadcasting, and NBC-Microsoft are examples of such deals. Nearly 20,000 alliances have been formed worldwide in the last two years, and of the 6,000 US alliances, more than half involved foreign partners. Reasons for such cooperation include: the reduction and sharing of risk; enhancing competitiveness by accessing core capabilities to enter new markets; leveraging complimentary asset positions; and the potential to increase market share. A survey of 5,000 US companies between 1987 and 1995 revealed that the average return on investment for strategic alliances was over 16 percent, which is higher than the industry's average.[9]

The world petroleum industry has traditionally been active through alliances. The old concessionaire system flourished, for example, through consortiums and the alliance between Chevron and Texaco in Caltex. The BP–Mobil partnership in European downstream, Amoco-Shell in the US upstream and the merger between Shell, Texaco and Saudi Refining to form the largest refining and retailing company in the US are the most recent examples of alliance activity in the petroleum industry. In the case of alliances between NOCs and major oil companies in the downstream sector, the ventures involved on the one hand could increase the vertical integration index of NOCs, while on the other hand, they may prevent major oil companies from pursuing such arrangements. It would appear that each venture or alliance between NOCs and major oil companies was carried out on its own merits and had a certain business background pertaining to the parties involved when the agreements were made. The two most important examples of this type are Star Enterprise, between Saudi Aramco and Texaco, and Unoven, between PDVSA and Unocal, the former being made at a time when Texaco was restructuring and seeking a new partnership after its settlement with Penzoil in the late 1980s.

Despite the advantages of downstream integration to oil producers, critics of the process do exist. Firstly, there is the rent-value-added tradeoff involved in the downstream reintegration. According to this view, investment funds directed by producers at international downstream integration reduces the availability of funds to expand upstream production capacity. Each crude barrel produced generates rent far in excess of the value added implied in the downstream investment. The risk associated with downstream investment, according to this view, is higher than the upstream risk, especially those in the low-cost, traditional producing areas of the Middle East and Venezuela. Moreover, at times of strong demand, much of the price increase in products sold on the world market is transferred to crude oil at the export source.

While investment in the upstream for rent seems easier and more straightforward in the short-run, it is no substitute for aggressive marketing to secure outlets for the extra crude in the longer-term. Since the sole crude producer can define a price and/or production strategy, the integrated producer can emphasize the value of the barrel of products. There need not be a contradiction between seeking rent or value added, and committing investment to the upstream and downstream. The experiences of Saudi Aramco, PDVSA, and KPC in expanding their upstream capacities and integrating downstream are cases in point.

The second criticism of international downstream integration by NOCs is related to the economics of investing in foreign rather than domestic refineries. The relevant comparisons here would be between building grass-root refineries in the producing country or acquiring existing facilities overseas, and between crude versus product-transportation economics. Environmental legislation and trade barriers are two further important considerations. Pure economics suggests acquiring refineries overseas rather than building domestic grass-root export refineries.[10] Idiosyncratic features pertinent to the situation of each country are also important. For example, Kuwait elected in the mid-1980s to construct sophisticated domestic refineries in order to refine its sour crude at times of declining world demand and widening differentials between sweet and sour crudes. And, when Saudi Aramco negotiated a 1990 downstream joint venture with a group of Japanese companies to build an export refinery in Saudi Arabia and acquire a share in an existing Japanese refinery, it was deemed then that the economics of the second option were preferable to those of the first.

The third criticism centers on the effect of such downstream investments on market stability and performance. It has been argued that since international vertical integration is not available to all oil producers because of differences in their resource endowments, a market situation might emerge where there would be traditional integrated companies, meganationals (involving the integrated NOCs), and non-integrated companies. This would result in the market being split between the interests of the haves and have-nots in refining, whereby the independent refiner and the independent producer would set the price of the incremental barrel. Accordingly, the market could be perceived by the oil producers that had embarked on downstream integration as being less than stable.[11]

The structure of the oil industry has consistently had an impact on prices and market conditions. Even in the halcyon days of the 1970s, when OPEC was in the ascendant, differences between high and low reserves existed. Such differences would continue, since resource endowments vary, and the structure of oil-producing industries differ.

The fourth misgiving about downstream integration by oil producers is related to the attractiveness of the investment itself. Refining has recently been losing money in most OECD countries, due to factors such as restrictive environmental standards and high capital and operating costs. Accordingly, NOCs could well pay higher prices for refineries that may require further injections of capital so as to meet new environmental standards.

The Future of Downstream Integration

Due to the reintegration process of the oil producers' oil industries, their status and ranking in the international industry has improved over the past few years. *Petroleum Intelligence Weekly*'s (*PIW*) ranking of the top ten international oil companies in 1995 listed five NOCs in their overall index, eight in the top upstream group and four in the top ten of the downstream companies. Three of the latter, namely Saudi Aramco, Pemex, and PDVSA have overseas downstream assets. (Table 2.3.)

Table 2.3 *Petroleum Intelligence Weekly*'s Ranking of Top 10 Companies Upstream and Downstream (1995)

Rank	Production (mb/d)		Refining Capacity (mb/d)		Product Sales (mb/d)		Overall Index	
1	S Aramco	8.574	Exxon	4.131	Shell	5.663	S Aramco	100.0
2	NIOC	3.721	Shell	3.535	Exxon	5.028	PDVSA	97.7
3	Pemex	3.137	Mobil	2.094	Mobil	3.075	Shell	97.0
4	CNPC	2.788	PDVSA	2.091	BP	3.002	NIOC	95.4
5	PDVSA	2.723	Chevron	1.836	PDVSA	2.903	Pemex	93.7
6	Shell	2.194	S Aramco	1.760	Texaco	2.352	Exxon	92.7
7	KPC	2.097	BP	1.745	Chevron	2.248	Mobil	84.5
8	Exxon	1.709	Petrobras	1.531	S Aramco	2.152	Chevron	78.2
9	LNOC	1.352	Texaco	1.453	Petrobras	1.657	KPC	77.6
10	Sonatrach	1.288	Pemex	1.357	Pemex	1.374	BP	76.6

Source: *Petroleum Intelligence Weekly*, December 18, 1995

There are differences in the integration index (refining to production) of the top ten companies in the *PIW*'s overall ranking. The five international majors have a combined refining capacity of 13.37mb/d and production of 7.015mb/d with an integration index of 1.91 (Shell 1.61, Exxon 2.42, Mobil 2.45, Chevron 1.88 and BP 1.38). Comparatively, the top five NOCs in *PIW*'s ranking have a combined refining capacity of 7.252mb/d and a crude production of 20.25mb/d with an integration index of 0.36 (Saudi Aramco 0.21, PDVSA 0.77, NIOC 0.29, Pemex 0.43 and KPC 0.45). The large divergence in ratios suggests there is room for further integration downstream by the producers, the extent and scope of which will depend on future market and investment outlook, and the strategies and options of the oil producers, as well as the international oil companies.[12]

The market environment in the medium- and long-term is characteristically different from the one that prevailed in the early years of reintegration. For example:

- Oil demand growth in OECD countries is projected to continue its slow downward trend while most of the growth in world oil demand is expected to be generated from the developing countries of Asia and Latin America.[13]
- Market relations are expected to expand to include states from the former Soviet Union. Economic liberalization in developing countries will also help open energy markets.
- Restrictive environmental legislation in OECD nations is expected to result in new demand patterns that will continue to affect the configuration and profitability of refineries.
- The increasingly competitive nature of the oil market in the upstream and downstream. The growing size and role of the forward and futures markets in oil and derivatives, and the active participation of oil companies, including NOCs, in such activities. This aspect of the market could affect corporate strategies and options.

While the aforementioned factors may not change the course of reintegration in the oil industry, they could pose many questions as to its future direction. From 1983 to 1995, out of 23 downstream deals concluded by oil producers, only two were in the Far East, while the rest were in the US and Europe. This could be due to the regulatory environment in those countries, which is changing rapidly. Recent deregulation of the petroleum sector in Japan involved gasoline price decontrol and the lifting of restrictions on product imports and service station construction. One could envision further downstream integration by the producers in those markets, considering their attractive margins. A closer look at the refining margins of Dubai crude in the Far East, compared to West Texas Intermediate (WTI) and Brent in the US Gulf Coast and North-West Europe respectively over the past eight years, shows a trend of commanding margins over other crudes in other markets. (Table 2.4.)

The second aspect of this new market environment is the growing energy needs of developing countries, especially for electricity generation. The World Bank projects the annual growth rate of demand for electricity in these countries to be approximately 5–7 percent. This entails increasing demand for fuels to generate the required capacity. One could envision fuel oil regaining some of its lost share in the electricity generation sector, which could be accomplished by forging alliances between oil producers and the power companies, or with the refiners of major developing countries.

Table 2.4 Refining Margins of Cracking Refineries in Three Main Markets (US$/bl)

	Brent NW Europe	WTI US Gulf Coast	Dubai Singapore
1988	1.82	2.60	2.36
1989	2.02	2.01	3.21
1990	3.01	2.64	4.43
1991	3.76	1.83	5.79
1992	1.94	1.22	3.52
1993	1.70	1.07	4.13
1994	1.60	1.24	2.97
1995	1.42	0.82	2.35

Source: *Oil Market Report*, Annual Supplement, IEA, September 1996

The third aspect of reintegration that is frequently referred to is some sort of downstream/upstream swap between the oil companies and the oil-producing nations. The new fiscal regimes in some oil-producing countries that allow foreign upstream investments on equity terms is seen as a bridge point whereby producers which were not able to integrate could use this vehicle to own downstream assets abroad.[14]

The extent of the use of this aspect depends on the fiscal/oil situation in each country. Some producers do not need foreign investments to increase upstream capacities, since they have access to technology and excess capacity in place. Others may not be inclined to pursue the downstream route because of their conviction that it is not as profitable as investing in the upstream.

However geographically diversified or intensive the reintegration process becomes, recent supply and demand projections predict an increased market share for Middle East producers. This means that their increase in downstream capacity abroad would be matched by an increase in demand for their crude.

Furthermore, the investment needed to increase or maintain existing capacities and increase downstream presence in overseas markets would be enormous, and might not be within some countries' financial or borrowing abilities, particularly considering the constraints their government budgets face. If this persists, it could slow down the pace of reintegration by oil producers, and downstream integration would be more selective across markets and investment opportunities.

CHAPTER 3

Strategic Options and Opportunities for Gulf Oil Companies in the Asian Market

Keiichi Yokobori

Introduction

Asia's energy demands, and in particular East Asia's, continue to grow. Given its population growth, industrialization, and urbanization, economic and energy research institutions predict that Asia's energy requirements will continue to rise well into the twenty-first century. This chapter examines current and future trends in Asian energy markets, describes potential strategic options and opportunities for Gulf oil companies, and highlights issues that need to be addressed if Gulf oil companies are to capture and capitalize on such market opportunities. Unless otherwise indicated, Asia covers five of the eighteen members of the Asia Pacific Economic Cooperation forum (APEC), namely Japan, China, Hong Kong, Republic of Korea, and Chinese Taipei, and members of the Association of South East Asian Nations (ASEAN), namely, Brunei, Indonesia, Laos, Malaysia, Myanmar, Philippines, Singapore, Thailand, and Vietnam. This excludes North Korea, Mongolia, and economies east of Bangladesh. The term "Asia-Pacific" indicates the region encompassing economies already defined as "Asia" as well as Australia and New Zealand.

Emerging Asian Energy Market

Over the past 25 years, Asian economies have produced some of the world's fastest growth rates, which in turn have contributed to the region's increased energy demands (Table 3.1). For example, China has recently registered a two-digit rate of annual economic growth and an annual ten percent increase in its energy demands. Between 1970 and 1995, the gross national product per head of South Korea rose almost elevenfold, Hong Kong fourfold and Thailand just over threefold. Even the Philippines, which has often lagged behind its Asian neighbors, has been producing respectable growth rates since 1993 (Table 3.2).

Table 3.1 Proved Reserves

	World		Europe		Asia Pacific	
	% to world	R/P ratio	% to world	R/P ratio	% to world	R/P ratio
Coal reserve	100	224	15.2	184	30.2	152
Oil	100	42.2	2	8.2	4.1	15.7
Natural gas	100	62.2	3.9	18.6	6.4	40.1

Source: BP *Statistical Review of World Energy* 1996

The growth in energy requirements also reflects the region's economic dynamism. Despite abundant coal resources and relatively large hydrocarbon reserves (Table 3.3), Asia, and in particular East Asia, is a net energy importer, and its external energy dependence is growing. Indeed, reliance on foreign energy sources is expected to continue. Japan's Ministry of International Trade and Industry (MITI), for example, released a report from its Advisory Committee for Energy, which predicts an increase in energy imports for the Asia APEC region from the 1992 level of 440.4 million tons of oil equivalent (mtoe) to 985.2mtoe by 2010. It anticipates that the total primary energy requirements in the region will almost double over the next 20 years, from 1,501.6mtoe to 3,009mtoe, while total primary energy production will increase by slightly less than twofold. While both non-oil production and demand will rise, oil requirements will expand at a higher rate. It is estimated that the demand for oil will increase from the 1992 level of 604.3mtoe to 982.7mtoe by 2010, and that domestic oil production will rise from 269.8mtoe recorded in 1992, to 302.3mtoe by 2010. Consequently, between 1992 and 2010, net oil imports will more than double, from 334.6mtoe to 680.4mtoe. Such a demand will further increase oil dependence on sources from outside the region, such as the Gulf (Table 3.4). The transition of China from an oil exporter to a net importer observed recently is symptomatic in this regard.

Table 3.2 Economic Growth Rate of Asian Economies (actual and projections)

	Annual 1992	1993	1994	Average 92-00	00-10	Source
China	13.2%	13.4%	11.5%	9.0%	7.0%	Energy Research Institute (Economic Social Development Plan)
Korea (R)	4.8%	4.7%	8.0%	6.6%	5.5%	Korea Energy Economic Institute
Taiwan	6.6%	6.2%	6.0%	6.5%	5.5%	Taiwan Electric Power
Hong Kong	5.3%	5.5%	5.6%	6.0%	5.0%	
Malaysia	7.8%	8.0%	8.8%	7.3%	5.8%	Malaysia Vision 2000
Thailand	7.6%	7.8%	8.3%	8.0%	6.5%	ASEAN 2010
Philippines	0.1%	1.7%	4.6%	6.0%	6.5%	Government Ecomomic Plan
Indonesia	6.4%	6.5%	6.8%	6.5%	7.0%	Repelita (Five-Year Plan)
Singapore	5.8%	9.9%	9.2%	6.0%	5.0%	Government's Strategic Economic Plan
Brunei	3.0%	2.0%	3.0%	3.2%	3.2%	Assuming the same rate of real oil-price increases
Japan	0.2%	0.3%	0.4%	3.0%	2.5%	Long-term outlook

Source: Japanese Ministry of International Trade and Industry (MITI) Advisory Committee for Energy (1995)

Asia's dependence on Gulf-produced oil has already been high (Table 3.4). According to recent British Petroleum (BP) statistics, Asia accounts for 58.1 percent of the Middle East's oil exports. Even in resource-rich Australia and China, it accounts for 45.3 percent and 24.4 percent, respectively, of their oil imports. At the same time, the importance of mutual interdependence in oil trade between the two regions will continue to expand if oil requirements as postulated here materialize in Asia, and if other oil-supply sources cannot readily absorb the incremental requirements.

While the aforementioned Japanese predictions are based on existing government economic and energy data, they appear conservative in comparison to the research carried out by Fereidun Fesharaki at Hawaii's East-West Center.[1] While Fesharaki's study includes Australia and India in the Asia-Pacific region, the exclusion of these two countries would make his base case comparable to MITI's findings. Oil demand projected for 2010 in the MITI outlook is 10 percent lower than the East-West Center estimates.

Table 3.3 Energy Outlook for Asia APEC Region

		Volume			Average Annual Growth Rate	
		1992	2000	2010	92-00	00-10
Production	Oil	269.8	276.2	302.3	0.3%	0.9%
	Coal	586.6	859.5	1,175.0	4.9%	3.2%
	Natural gas	91.8	157.4	209.6	7.0%	2.9%
	Others	113.0	181.7	336.9	6.1%	6.4%
	Total	**1,061.2**	**1,474.8**	**2,023.8**	**4.2%**	**3.2%**
Demand	Oil	604.3	744.1	982.7	2.6%	2.8%
	Coal	683.4	957.3	1,403.9	4.3%	3.9%
	Natural gas	100.9	174.8	285.4	7.1%	5.0%
	Others	112.9	181.7	336.9	6.1%	6.4%
	Total	**1,501.6**	**2,057.9**	**3,009.0**	**4.0%**	**3.9%**
Net imports required	Oil	334.6	467.9	680.4		
	Coal	96.8	97.8	228.9		
	Natural gas	9.1	17.4	75.8		
	Total	**440.4**	**583.1**	**985.2**		

Source: MITI Advisory Committee for Energy (1995)

Table 3.4 Middle-East Dependence (1995)

	Australia	China	Japan	Other Asia	Total Asia
Oil Imports from Middle East (mtoe)	11.1	7.7	206.8	253.8	**479.4**
% to Total Oil Imports	45.3	24.4	74.9	80.1	**73.9**
% to Middle East Oil Exports	1.3	0.9	25.1	30.8	**58.1**

Source: BP *Statistical Review of World Energy* 1996

Furthermore, the latter also postulates high oil-demand growth for India, an annual growth rate of 5 percent for the period 1993–2000 and 3.8 percent for the period 2000–2010, an average of 4.3 percent over a 17-year period (Table 3.5). The variance in data arises from different assessments of energy-efficiency improvements, availability, and competitiveness of alternative fuels and supply sources, including such factors as the timing of planned projects' realization, and other underlying economic trends. Such issues will be revisited later in discussing the opportunities for Gulf oil producers.

Table 3.5 Comparison of Oil Demand Outlooks (mtoe)

		1992	1993	2000	2010
MITI		604	na	744	983
EWC	Base Case Asia Pacific	na	703	931	1,250
EWC	Asia Pacific (without Australia and India	na	608	806	1,077

Note
Converted at 49.5 toe per b/d and with 5% loss for conversion from crude oil to products

Sources: MITI Advisory Committee for Energy
Fesharaki, F. et al., *Pacific Energy Outlook: Strategies and Policy Imperatives to 2010* (Honolulu HI, East-West Center, 1995)

Opportunities for Gulf Oil Companies

Asia's high level of dependence on Middle East hydrocarbon supplies would suggest good market opportunities for Gulf oil suppliers. Whether market opportunities for such companies could be further expanded would, however, depend on several factors. These include: the prospect for the continuous expansion of Asian economies; the growth patterns of differing economic sectors, e.g. transportation; the rise in energy-efficient trends; competition from alternative oil suppliers; natural gas (particularly from non-Gulf sources); nuclear-based electricity; and coal. Competition can be viewed not only in terms of price, but also in terms of environmental impact. These factors will be examined later.

The prospects for continued Asian economic expansion appear to be shared by economic observers in general. They are supported not only by the region's recent dynamic activities, but also by the trend in governmental policy towards deregulation. Regional cooperation through such forums as APEC could contribute to continued economic development and increased transactions of goods, services, technologies, and information among economies inside and outside the region. While the extent to which such deregulation would contribute to further economic growth remains a subject that future studies should address, a reasonable degree of optimism appears to prevail at present. However, this does not ensure that targets outlined in "national plans" will materialize as intended.

Table 3.6 Final Energy Consumption in 10 Asian APEC Economies[1] (mtoe)

	1980	1992	2000	2010	1992/1980	2010/1992
Industry	256.0	450.7	636.3	893.3	4.8%	3.9%
	(63.2)	(61.0)	(58.6)	(53.2)		
Transportation	43.4	93.3	166.9	357.9	6.6%	7.8%
	(10.7)	(12.6)	(15.4)	(21.3)		
Residential/Commercial	105.7	195.4	283.4	426.6	5.3%	4.4%
	(26.1)	(26.4)	(26.1)	(25.4)		
Total	405.1	739.4	1,086.6	1,677.8	5.1%	4.7%
	(100)	(100)	(100)	(100)		

Note
1 Excluding Japan

Source: Institute of Energy Economics, Japan

The regional energy-demand pattern will vary (Table 3.6). For example, in the region's developing economies, industry's share, which has accounted for over 60 percent of total final-energy consumption (TFC), will decline, although in some countries the growth of energy-intensive industries could cause higher industrial-energy consumption. Industrial energy-use efficiency is generally assumed to improve with technological advancements. Indeed, it has been argued that latecomers can take advantage of the most recently developed technology, and thus leap-frog to the most energy-efficient technologies.[2] The extent to which developing countries could leap-frog in specific circumstances again remains a subject of more careful study. At any rate, for these reasons, the industrial sector's energy use will increase at a slower pace than overall energy consumption growth.

On the other hand, the transportation sector's energy demand, which accounts for around 13 percent of TFC, is expected to grow faster than any other sector. The high growth of transportation demand will be characterized by highway transport and aviation, accompanied by weak or deteriorating rail transport. This is already taking place, not only in industrialized economies, but also in developing ones. For example, Chinese passenger-rail-transport volume has shown sluggish growth, while highway transport and civil aviation volume have registered steady increases. The OECD experience of a growing demand for "higher grade" vehicles that offset vehicle fuel efficiency could be repeated in developing Asia.

The residential and commercial energy use will retain its present share of approximately 25 percent. Oil demand in this sector is also dependent upon many factors besides overall sectoral energy use, such as shifts from traditional fuel woods and other biomass to fuel oil, and competition with gas (derived or natural) and electricity. While the electricity generation sector's dependence on coal is likely to continue, its oil use in the region as a whole will not increase in a similar manner, due to the anticipated reduction in Japan, which will to some extent offset the rise in oil use in other economies. This largely results from increased nuclear-power generation, according to the MITI projections. Even if nuclear expansion does not materialize, LNG or coal could be the preferred options, although some believe that oil would be the easiest replacement. Even in this case, the increase in oil use would be marginal, as no new oil-fired power plants will be built (Table 3.7). Coal and nuclear power are almost completely confined to the generation of electricity, and are promoted by ambitious government plans in Asia. Delays in their projects could result in increased oil use, if that market is not captured by natural gas. However, relative abundance of heavy fuel may not make transport-related energy demand a determining factor of oil-demand growth, as seen later.

Thus, the oil-intensive transportation sector's growth will lead to higher oil demand.

Table 3.7 Electricity Generation by Fuel in Asia APEC Region (10^9kWh)

	1992		2000		2010		Additions 92-00		Additions 00-10	
	Total	Excl. Japan	Total	Excl. Japan	Total	Excl. Japan	Total	Excl. Japan	Total	Excl. Japan
Hydro	245	162	389	293	665	538	144	131	276	245
Nuclear	313	91	449	141	879	401	136	50	430	260
Coal	770	683	1,475	1,325	2,919	2,754	705	642	1,444	1,429
	409	190	376	225	415	305	-33	35	39	80
Natural Gas	235	59	466	230	724	491	231	171	258	261
Geothermal and others	9	7	28	23	49	29	19	16	21	6
Total	**1,981**	**1,192**	**3,182**	**2,237**	**5,651**	**4,518**	**1,201**	**1,045**	**2,469**	**2,281**

Source: Calculated by the author on the basis of the support material for the deliberations of the MITI Advisory Committee for Energy (1995)

Table 3.8 Asia APEC Oil-Product Demand

	Volume	(Mtoe)		Annual growth	
	1992	2000	2010	92-00	00-10
Naphtha	59.8	68.3	97.6	1.7%	3.6%
Gasoline	88.7	139.3	218.8	5.8%	4.6%
Kerosene	55.4	71.3	95.9	3.2%	3.0%
Diesel	127.4	170.0	234.8	3.7%	3.3%
Heavy fuels	160.0	181.3	203.2	1.6%	1.1%
Others	75.5	82.9	104.4	1.2%	3.0%
Total	**366.8**	**713.1**	**954.7**	**2.9%**	**3.0%**

Source: MITI Advisory Committee for Energy (1995)

The MITI Advisory Committee's report anticipates that the region's gasoline demand will increase from 88.7mtoe (1992) to 139.3mtoe (2000), an annual increase of 5.8 percent; and then to 218.8mtoe (2010), a 4.6 percent annual increase. Heavy fuels would increase at a slower rate, i.e. 1.6 percent per annum (except 1.2 percent per annum for "others") from 1992 to 2000, and at a rate of 1.1 percent per annum from 2000 to

2010. These figures can be compared with the average increase in annual petroleum products demand of 2.9 percent by 2000 and 3.0 percent by 2010 (Table 3.8). Consequently, gasoline, which was the third-largest fuel, accounting for 15.6 percent of total oil-product demand in 1992, will be the second-largest by 2010, accounting for 22.9 percent, followed by 24.6 percent for diesel, which ranked second with 22.4 percent in 1992. Heavy fuels, which were the largest in 1992, with 28.2 percent, would be ranked third with 21.3 percent.

With regard to competition from other fuels, as highlighted above, by 2010, the heavy end of products will face challenges from alternative electricity-generating fuels such as coal, nuclear power and natural gas. Coal's abundance in the region, despite its more polluting characteristics certainly represents for many — in particular the Chinese — a convenient option. It should be noted that Asia is known for its aggressive nuclear-power plans. While the public has concerns regarding radioactive safety and the potential for weapon proliferation, nuclear power's lack of air-pollutant emissions such as sulfur and carbon is attractive to the region's policy makers. While coal and nuclear power would compete with oil in relatively limited segments of the market, such as electricity generation and some industrial use, delays or other failures in implementing such ambitious plans and programs may well produce an additional, though limited, opportunity for oil.

Natural gas also offers relatively clean attributes in emitting pollutants in combustion. Integrated combined-cycle technologies allow short lead time for power-plant construction and ease of operation, which in the current economic and business climate are aspects increasingly appealing to energy-industry executives. Natural gas from non-Gulf regions could also become a more serious competitor. For example, such competition would arise from the potential supply of natural gas and oil from Siberia and Central Asia, as well as from other untapped indigenous sources in South-East Asia. The development of a pipeline crossing Mongolia from the Kovyktinskoye gas field has been agreed upon between China and Russia, and has attracted substantial Korean and Japanese interest. The development of Central Asian hydrocarbon resources, for example, plans to pump gas from Turkmenistan and Russia to China and then onward to Korea and Japan, and the construction of a pipeline traversing Afghanistan and the Indian subcontinent have been examined. Although there remains a multitude of political, economic, and technological issues to be resolved on these projects, they could certainly affect the future of the Asian energy-market scene.

At the same time, crude oil imports from sources such as West Africa are also rising, and are expected to continue (Fig. 3.1). However, current oil price levels, particularly those of Gulf crude, appear to be preventing West African exports from making any significant impression on the Asian oil market. In fiscal 1995, the average CIF crude oil price from Gabon was 6,506 yen/kl against the average of all imports of 11,004 yen/kl. However, in fiscal 1996, it was priced 15,895 yen/kl, against the import average of 15,186 yen/kl. The import volume of African crude was reduced from 1.8bn liters in fiscal 1995 to 1.2bn liters in fiscal 1996. Middle East oil was priced at approximately the import average.

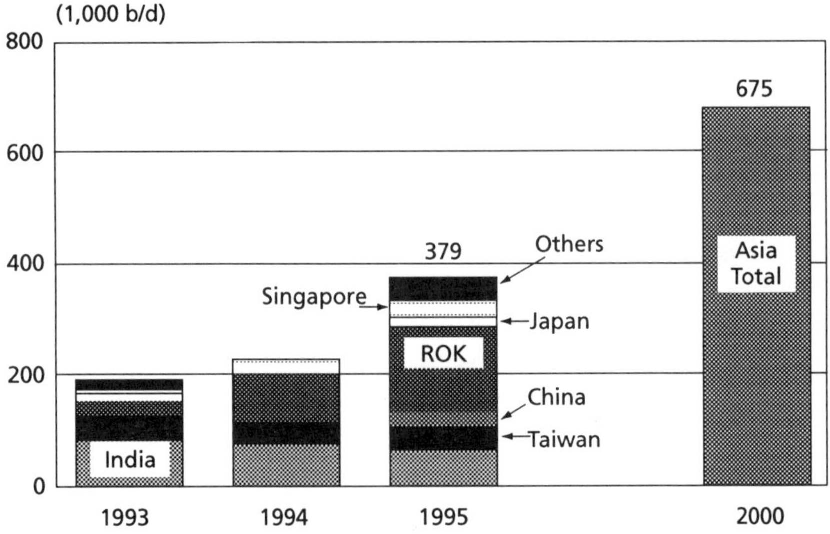

Source: Yong Ho Jang and Takashi Kato (1996)

Fig. 3.1 Crude Oil Exports from West Africa to Asia (actual and prospective)

Among fossil fuels, oil has, besides its ease in handling (storage, transport, etc.) a relative environmental advantage over coal because of its lower sulfur oxide (SOX) and carbon dioxide (CO_2) emissions. Government agencies and energy suppliers in many developing Asian economies have had a low level of environmental concern in the past, which has prevented further shifts towards lower-pollution fuels such as natural gas. This could allow fuel competition to be conducted in the economic arena without necessarily taking account of external costs, such as environmental pollution prevention and abatement. However, this situation may alter in the future.

Challenges for Gulf Oil Companies

While prices will play an important role, it will not necessarily be a comparison of price against calorific value that counts. Consumers will take into account other factors, such as environmental costs associated with its eventual use, reliability of supply, including logistic safety, price transparency, and so forth. Such issues have to be addressed by Gulf oil companies if they are to capture and expand market opportunities in Asia.

Firstly, energy-related environmental concerns continue to grow in the region; many cities have already experienced traffic congestion and a polluted atmosphere caused by increased use of fossil fuels. In particular, the production of SOX and nitrous oxide (NOX) has generated direct and immediate health hazards, especially in the region's congested urban areas. According to the United Nations' Population Division, 359 million people lived in Asian cities with populations over 1 million in 1990, and by 2015, this figure is expected to reach 903 million (Table 3.9). With such further concentrations of population, urban atmospheric pollution levels will soar. These pollutants also produce acid rain, which damage the environment more generally. The increased use of flue-gas desulfurization equipment and other pollution-abatement techniques and technologies should be encouraged. While these methods tend to lower energy efficiency initially, and increase investment and operating costs, technological progress will in time reduce such costs.[3]

Some simpler and cheaper, though less effective, abatement technologies are also being developed. For example, the Japanese are researching and testing simplified flue-gas desulfurization equipment, which achieves 80 percent of the performance at half the cost of a more complex system. Fossil fuels produce in their combustion CO_2, one of the major greenhouse gases that causes climate change. In light of the lack of any economic and technologically feasible carbon sequestration from flue-gas, CO_2 emissions could be reduced through more efficient energy-utilization technologies, and by the use of lower-carbon-content fuels, such as natural gas and nuclear power. While both options would also contribute to SOX and NOX emissions, more efficient use of finite fossil fuels would be mutually beneficial to oil suppliers and consumers. With the exception of CO_2 emission controls, some of which could incur very high costs, pollutant emission controls could be implemented now at reasonable costs.

Table 3.9 Population in Cities with Over One Million Residents, by Region, 1950–2015 (population in millions)

Region	1950	1970	1990	2015
Africa	3	16	59	225
Latin America	17	57	118	225
Asia	58	168	359	903
Europe	73	116	141	156
North America	40	78	105	148

Source: United Nations (UN) Population Division, *World Population*

The failure to consider environmental concerns, namely to pay such external costs or to take safeguard actions, may not only prevent these fuels from continuing to contribute to our welfare, but also eventually cause oil to lose its economic advantages over other fuels. Adverse environmental risks associated with oil, such as marine pollution, need continuous preventive action. In short, while uncertainties over climate change and greenhouse gas emission controls still remain, it becomes reasonable to bear the costs of relevant environmental protection in the use and supply of energy (and goods and services). It should be pointed out that almost all energy sources will create accompanying environmental hazards in their production, transport, transformation, or use. The integration of such externalities would also contribute to a fairer choice of fuel, which would obviously have an impact on the competitiveness of oil.

Secondly, the supply of energy resources from distant sources inevitably raises the issue of reliability of supply. The almost exclusive dependence on tanker transport of oil, which differentiates the Asian market from the North American and European markets where at least intracontinental pipeline networks are established (though limited in Europe), could imply vulnerability of oil supply from the Gulf. For example, the increased passage of tankers through narrow straits near Indonesia, such as Lombok, Malacca, and Sunda, has already raised concerns. Indeed, the October 1997 collision between the Cypriot-registered tanker *Evoikos* and the Thai supertanker *Orapin Global*, and the subsequent spillage of 25,000 tons of marine fuel oil 13 kilometers south of Singapore's main island highlights such regional concerns. There were worries that such a spill would force the Maritime and Port Authority of Singapore (MPA) to close sea lanes connecting the Strait of Malacca and the South China Sea, thereby disrupting one of the world's busiest ports and oil-refining centers. As experienced in past oil-supply disruptions, the supply shortage may occur unevenly among differing product groups. This will call for more care in maintaining flexibility in product mix, which must be considered in future refinery-capacity expansion needs. Furthermore, the aforementioned inclination towards lighter products and anticipated tighter environmental requirements in Asia would induce customers to select a lighter crude oil with a reduced sulfur content. Gulf oil producers could both mitigate Asian consumer concerns and achieve their preference by participating in logistic infrastructure building projects, such as pipelines, refineries, or storage facilities somewhere in between. Such a prospective cooperative venture has been floated by Asian experts.[4] At the same time, customers could also share producers' interests in enlarging the supply potential. In this sense, the participation of Japanese companies in developing hydrocarbon resources, such as the Japan Oil Development Company's involvement in offshore Abu Dhabi fields, has contributed to increased global supply. Thus, increased supply stability would serve both suppliers and consumers.

Thirdly, the price of oil also counts. At present, Asian oil trade is generally linked to the price of Dubai crude. This is contrasted with oil transactions with Europe, which are based on Brent, and with America, based on West Texas Intermediate (WTI), both being indigenous to the customers' market. While it is inevitable that a Gulf crude oil will be used as a marker due to the lack of an alternative tradable Asian crude oil for reference, several attributes of this particular crude cause problems. Firstly, production has declined from 350 thousand barrels per day (tb/d) in the first quarter of 1993 to 260tb/d in the second quarter of 1996, with a further decline to 214tb/d in 1997. Secondly, its main customer, India, is increasing its demand for oil, which has resulted in the reduced availability of Dubai crude to the open market. This marginal nature, together with Asia's growing oil demand, would strengthen its price beyond what the underlying demand-and-supply balance could justify. Thus, if compared with Europe- and North America-bound Arabian Light crude oil (AL), the Asia-bound AL formula price, which uses Dubai as a marker crude, has been kept US$1 higher per barrel since 1992 (Fig. 3.2). The fact that it tends to be priced higher than the Brent or WTI-based formula price produces concern among Asian customers. At the same time, in North America and Europe, despite the short-term fluctuations, the difference between the spot price and the netback value of marker crude remains narrow in the long run (Fig. 3.3). In Asia, this difference shows more marked gaps (Fig. 3.4). This suggests a weaker linkage between crude and products in Asian markets, and less-transparent market conditions. While such market rigidity could be gradually reduced with growing deregulation and wider use of forward and futures markets of both crude and products, the dependence on Dubai crude, with its declining production, and growing Asian oil demand, especially for lower sulfur and lighter crude, could result in higher and more volatile crude prices. Indeed, more volatile product markets resulting from greater

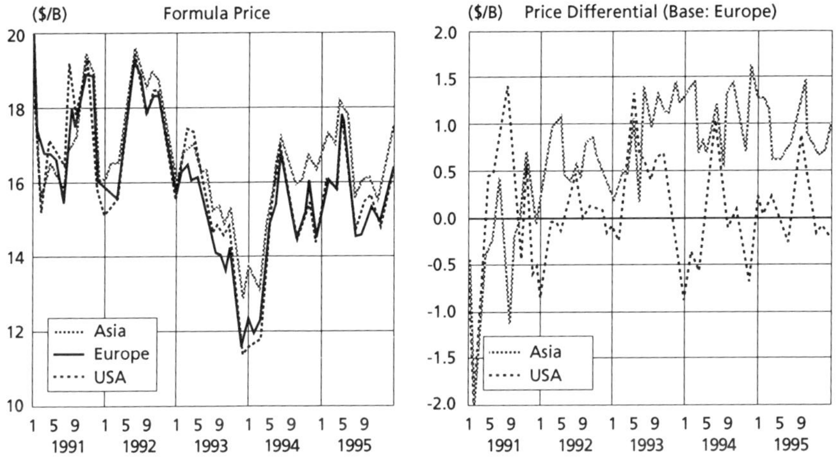

Source: Yong Ho Jang and Takashi Kato (1996)

Fig. 3.2 Trends in Market Crude Spot Prices and in Price Differentials with Netback Value

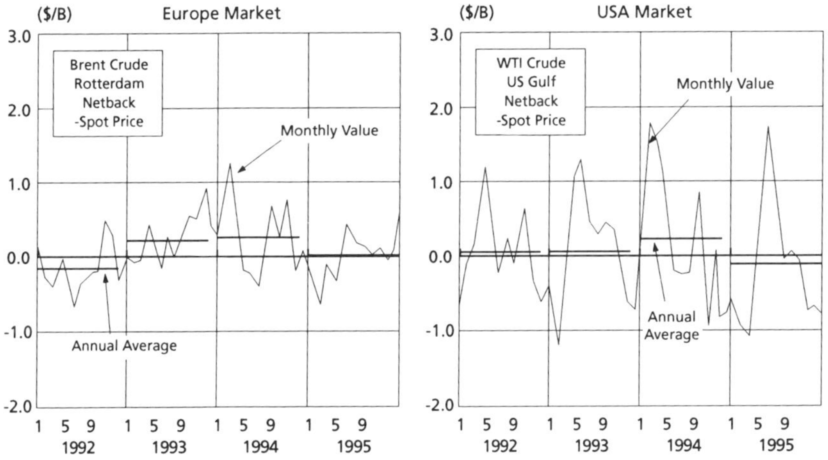

Source: Yong Ho Jang and Takashi Kato (1996)

Fig. 3.3 Trends in Market Crude Spot Prices and in Price Differentials with Netback Value competitive pressure and higher crude costs, could be seen as an obstacle to developing a healthy oil industry in the region. In order to avoid such market instability, the search for an alternative acceptable marker-crude or pricing formulae should be encouraged.[5]

This peculiarity of the Asian oil market, with the virtual lack of competing indigenous oil — except for Indonesia and China, whose oil market has not yet been fully opened to the world — could result in both suppliers and customers becoming indifferent to each other's perceptions and interests. If suppliers understood the needs of consumers better and were convinced of the economic benefits which closer ties would bring, perhaps then a stronger commitment to explore a mutually acceptable solution would emerge. Conversely, energy consumers may be able to understand the producer's preoccupations better by participating in upstream operations.

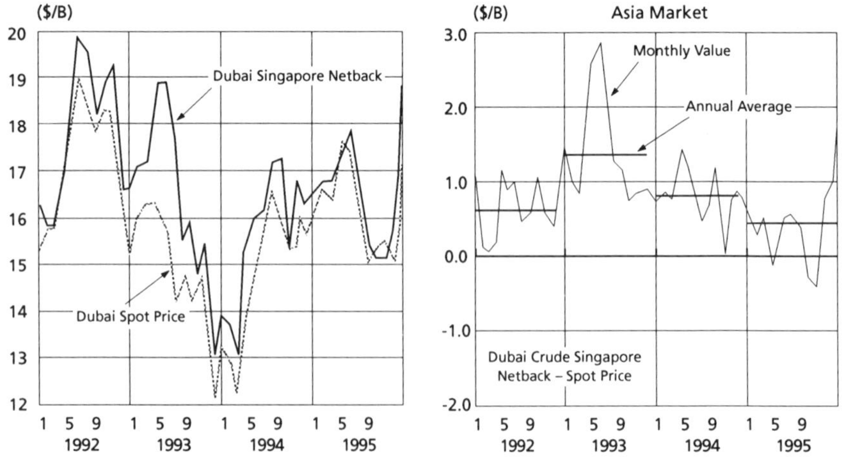

Source: Yong Ho Jang and Takashi Kato (1996)

Fig. 3.4 Trends in Market Crude Spot Prices and in Price Differentials with Netback Value

It is true that many Asian oil and energy markets have been regulated and closed to outside investors. It should also be noted that Asia is composed of diverse economic, social, and political zones. However, Asian energy markets are now moving towards more deregulated systems. For example, the inaugural APEC Energy Ministers' meeting held in Sydney in August 1996 endorsed 14 non-binding energy-policy principles which encourage the removal and reduction of impediments to free capital, technology, and information aimed at more efficient energy use and environmentally sound energy systems. They also encourage open energy markets and subsidy removals. Progress will be monitored periodically; although the situation will not change overnight, Asian markets are opening up. In this sense, it is encouraging to see that Gulf oil companies are now investing or forming joint ventures in downstream segments in many parts of Asia (Table 3.10). At the same time, customers should be prepared to share mutual interests with suppliers in a broad economic and social sense. Cooperation could, as suggested, be extended beyond building an energy-related logistical infrastructure and include the exchange of information, technology, and other resources.

There will certainly be wider and more important economic implications involved in Gulf oil companies forging a stronger relationship of mutual prosperity with Asian customers. Initiatives could include common action on addressing environmental issues as outlined above; industrial cooperation could also be broadened. For example, energy efficiency improvements and environmental abatement technologies would contribute to the longevity of finite fossil fuel reserves. Advanced and more economic flue-gas desulfurization and dinitrification technologies could enhance oil's market potential over other fuels. Indeed, carbon sequestration technologies, which are far from marketable today, could eventually address the problem of climate change arising from the use of fossil fuel.

In essence, meeting these challenges would create options for Gulf oil companies to take advantage of opportunities in the Asian market.

Table 3.10 Gulf Oil Companies' Participation in Asian Downstream Projects (as of January 1996)

Oil-producing countries	Economies of Destination	Date	Description of Projects	Share (%)	Note
Kuwait	Thailand	1/90	Construction of a gasoline station network with 200 outlets		A target
	Thailand	10/95	Joint refinery with TPI (Revong: 300tb/d)	30.0	
	Singapore	10/90	Participation in Singapore Petroleum Co. Ltd	10.6	
	Pakistan	8/95	Joint refinery with Pakistan State Oil Co. Ltd (120tb/d)		
	India	9/95	Joint refinery with IOC (120,000 b/d)	26.0	FS from Oct '95; Total costs US$1.7bn
	India	9/95	Cochin refinery (200tb/d)		Under study
Saudi Arabia	Korea	7/91	Ong-Zuang refinery with Ssangyong (150tb/d)	35.0	
	Philippines	12/93	Stock acquisition Petron Corp., a refining and product sales affiliate for PNO	40.0	
	China	1995	Construction of a new refinery (200tb/d)	45.0	Ssangyong shares 15%, to be completed in 2000
	Taiwan	1995	Participation in CPC	30.0	Under study
Abu Dhabi	Pakistan	1974	Pak-Arab refinery company (PARCO)	40.0	Pipeline in operation
	Pakistan	2/94	Construction of a new refinery (100tb/d)		
Iran	Pakistan	5/91	Construction of a new refinery in Karachi (120tb/d)	50.0	
	China	7/93	Construction of a refinery		
Oman	Thailand	4/93	Refinery (130tb/d)	64.0	A joint venture with Caltex and PAT
	India	3/93	Joint refinery with BPCL (120tb/d)	28.0	
	India	3/93	Joint refinery with Hindustan Petroleum (120tb/d)	28.0	

Source: Idemitsu Kousan

Conclusion

How to capture opportunities offered by the maturing Asian energy market will pose challenges for Gulf oil companies. One way of coping with such challenges would be to meet the needs of Asian consumers. This would involve delivering a stable and adequate oil supply at affordable prices. In turn, Asian customers should understand what "stable and adequate oil supplies at affordable prices" would mean to Gulf producers, i.e. open and fair access to customers, with a fair return. Such a common understanding could be fostered by providing a safe and secure logistical infrastructure for the distant consumer, the reciprocal participation in both upstream and downstream operations to cement mutual business interests, the development of a transparent pricing scheme, greater exchange of goods, services, and technologies in a broader context, and appropriate actions to safeguard the environment. Coupled with energy deregulation and increased market-conforming policies in the Asian markets, such options would certainly enhance the access of Gulf oil companies to the expanding Asian market.

CHAPTER 4

Oil Refining and Petroleum-Product Specifications in Asia

Tilak Doshi

Introduction: The Clean Fuels Imperative in the Cities of Asia

It has been frequently noted in oil industry circles that the sheer rapidity of change in petroleum-product specifications in the Asia-Pacific region during the first half of this decade has caught most observers by surprise. However, it is not often asked just why this is so. It may be worthwhile to spend a little time on this question before proceeding to characterize the nature of product specification changes in Asia Pacific, and the impact they have had, and will continue to have for the forseeable future, on the region's refiners.

As the twentieth century draws to a close, public opinion in OECD countries has shown much concern with environmental issues. Indeed, industry is increasingly compelled to consider the impact of its activity on the environment. For developing countries, it has often been assumed, not without good reason, that there is no rational reason why they should choose between economic growth and environmental quality in the same manner as the OECD countries; that is, if there is indeed a trade-off between growth and the state of the environment. More specifically, it has been implicitly assumed that the demand for environmental quality is essentially a function of increasing average incomes.

With hindsight, it has become evident that the demand for better environmental conditions in many Asian countries has been more directly related to the size and political weight of the urban classes in these countries. The concentration of the middle and professional classes in Asia's booming (and often grossly polluted) cities — and their increasing role in the political arena of Asian government, arising from the liberalizing policy initiatives adopted in many of the countries in the region — have led to demands for many of the environmental standards that are taken for granted in the richer countries of the OECD.

With respect to petroleum product specifications in the Asia-Pacific area, a vital characteristic is the extreme heterogeneity of the region. Few generalizations apply across the region; this is also true of product specifications. The first question that naturally arises is whether Asia Pacific can be said to constitute a "region" in any meaningful sense at all. For the petroleum analyst, Asia Pacific — or in some ways the more appropriate "East-of-Suez" — is considered as a region for the following three reasons:

- It has been the engine of global demand growth since 1986, on a scale that differentiates it from other regions;
- With limited domestic crude petroleum resources, the region is a major destination of the world's crude exports, primarily from the Middle East;
- It is the only major consuming region in the world to have a refining-capacity shortage (although with the spurt in base-distillation capacity coming onstream between 1994 and 1997, the capacity deficit has been temporarily attenuated), and hence is the destination of surplus petroleum products from all parts of the globe.

The heterogeneity of the East-of-Suez region, with great economic and technological diversity among nations (and consequently with a range of divergent demands on the region's petroleum industry), has led to a constantly shifting constellation of intrasectoral competition and complementarity. For example, there is a considerable two-way trade in diesel and fuel oil among countries in the region, due to specification differences. Over the past few years, the instances of simultaneous surplus and deficit in different grades of the same product in East-of-Suez markets have multiplied, and at times, market prices of specific-grade products have moved in directions contrary to perceived fundamentals. Indeed shifting product differentials, and their implications for arbitrage and logistics have become a primary characteristic of the petroleum product trade in East-of-Suez markets. In short, rapid changes in product specifications in the Asia-Pacific area have emerged as a major challenge to refiners, both those from within the region and those from other regions that frequently place cargoes in the regional market.

Asia and Structural Changes in Global Oil Demand

Unlike other energy forms, oil is a global business. The world oil market is important to the international economy simply by virtue of its size. Oil accounts for approximately one tenth of the value of world trade, a far greater share than any other single commodity. While coal and gas are also internationally traded, they are proportionately small relative to oil. While the world crude oil market exhibits the characteristics of a global market,[1] trading in coal and natural gas — via pipelines or as LNG — is either regional in nature or restricted to a concentrated group of buyers and sellers as in the example of the LNG trade.

Oil is also the most intensively used energy form. According to the International Energy Agency (IEA), oil accounted for over 40 percent of global primary-energy demand in 1993; coal, its nearest competitor, accounted for 27.5 percent, followed by gas with 22 percent, and nuclear and all other forms constituting the rest.[2] In 1971, oil represented almost 49 percent of total primary-energy demand.[3] The decline in oil's market share since then, to some 40 percent currently, can be attributed to the oil price shocks of the 1970s, which led to fuel substitution, increased energy efficiency, and conservation through the 1980s.

The oil industry is undergoing a shift in the geographical distribution of oil-product consumption, a process that has intensified since the latter half of the 1980s. According to the IEA (Table 4.1), the developed countries of North America and Western Europe accounted for almost 60 percent of world oil demand in 1971; this has now decreased to less than 50 percent, and this trend is likely to continue.[4] Strong, persistent economic growth, combined with the demographic weight, in the emerging market economies outside North America and Western Europe, particularly in East and South-East Asia, will mean that they account for a larger share of world oil demand. For example, by the year 2010, countries outside North America and Western Europe are expected to account for almost 60 percent of all oil consumed.

Table 4.1 Regional Distribution of World Oil Demand (percent)

Geographic Area	1971	1993	2010
North America	33	29	25
Western Europe	26	19	16
Sub-Total	**59**	**48**	**41**
FSU/E Europe	14	11	9
Asia/Pacific	15	24	30
Other	12	17	20
Sub-Total	**41**	**52**	**59**

Source: *World Energy Outlook*, IEA, 1996

In the fifteen-year period 1980–95, when the developing countries of the world (including China) accounted for almost 10 million barrels per day (mb/d) of incremental oil demand, the OECD accounted for only 1.4mb/d, while the former Soviet Union (FSU) showed a decline of 4.1mb/d (Fig. 4.1). In-house projections (Fig. 4.1), utilizing consensual long-run economic and demographic growth-rate forecasts that can be taken as representative of the oil industry's outlook, suggest that over 70 percent of expected global incremental oil demand (of some 28mb/d) over the following 15-year period, 1995–2010, will be accounted for by developing (non-OECD, non-FSU) countries. China alone will account for more than 10 percent of global incremental demand over the fifteen-year period to 2010. A more detailed forecast for the ten-year period 1996–2005 (Fig. 4.2) indicates the importance of non-OECD Asia; the region is expected to account for half of the global incremental petroleum-product demand.

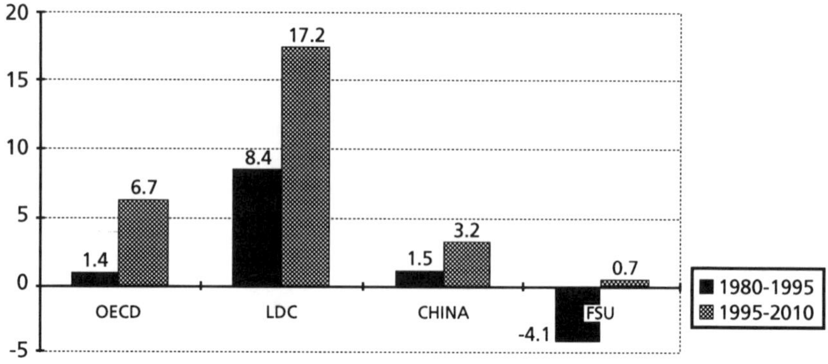

Source: Economics and Business Environment Group, ARCO Corporate Planning

Fig. 4.1 Incremental Oil Consumption by Region (mb/d)

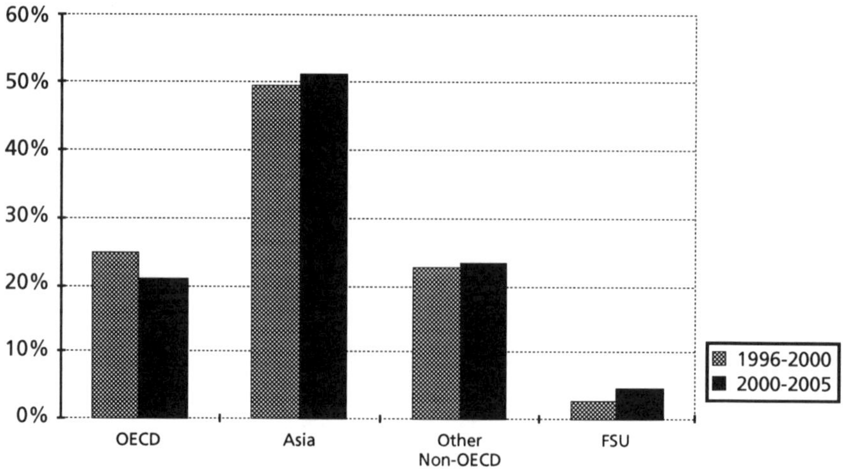

Source: Economics and Business Environment Group, ARCO Corporate Planning

Fig. 4.2 Percent Distribution of Incremental Product Demand

Besides the expected geographical distribution of future oil consumption in favor of the emerging market economies, particularly those in Asia, the most important structural change on the demand side is the concentration of oil use in specific markets, where substitution by another energy source is currently not possible or extremely costly, namely transport. The use of oil for petrochemical feedstocks is still a relatively limited phenomenon, though increasingly a high value-added activity in many cases. Transportation will increasingly claim the lion's share of oil use, with heating fuel's share dropping progressively. In many countries, particularly the OECD, oil has lost much of its share of boiler-fuel markets to natural gas, and is no longer the preferred fuel for electricity generation, which is seen increasingly as the domain of natural gas and coal.

Associated with this structural change in sectoral oil demand is a changing distribution of demand, by type of refined product, which now increasingly tends towards the light and middle distillates at the expense of heavy fuel oil. (Fig. 4.3.) While this change has occurred throughout the global market, it is most accentuated in the OECD countries, where over the past decade, the share of fuel oil has dropped, primarily as a result of having been displaced from the electricity-generation sector. This structural shift will continue into the next century in non-OECD regions. However, there is some uncertainty as to the pace of change, primarily as it is unclear whether the tremendous latent demand for electricity can be fully met by natural gas and coal.

The development of markets for middle distillates is mainly based on the sharp rise in aviation and highway transport, and on the increased use of diesel for commercial vehicles, particularly in the developing countries of Asia. The bias towards the middle of the barrel is apparent in industry forecasts of product demand in non-OECD Asia. (Fig. 4.4.)

In many parts of Asia, the skewing of demand in favor of middle distillates has been accentuated by government regulations.[5] The justification for subsidizing diesel as an "industrial" fuel has often been made in terms of protecting comparative advantage in the manufacturing export sector. Many Asian countries also subsidize kerosene on social-welfare grounds when the fuel is used by poorer rural households for cooking and lighting. Often these subsidies have been made possible by steep taxes on gasoline as a "luxury" fuel. Countries such as India and Indonesia have thus developed distorted consumption patterns as a result of government pricing policies.

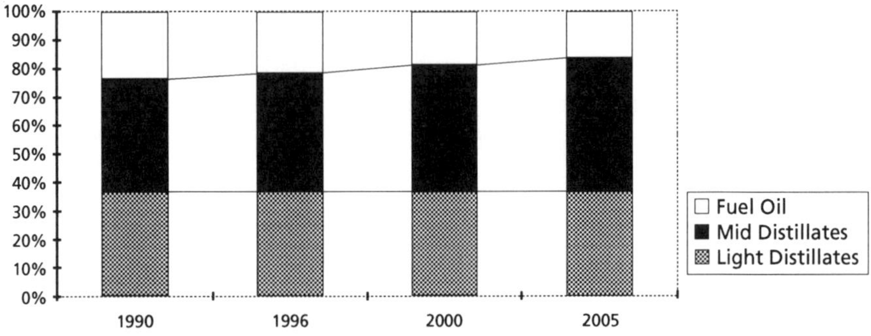

Source: Economics and Business Environment Group, ARCO Corporate Planning

Fig. 4.3 Global Product Demand (percent distribution)

Petroleum Product Specifications in Asia Pacific

In the foregoing, the major characteristics and trends of the global oil market on the demand side, with emphasis on Asia, were established. Among the major features noted were the structural shift in oil use towards the middle and light distillates, and the geographical importance of the emerging-market economies, particularly those of Asia. With respect to environmental requirements of Asia's downstream oil industry, two main points were emphasized. Firstly, the pressures on Asian governments to mandate increasingly stringent standards for petroleum products should not be underestimated: rapid economic growth is almost always accompanied by assertive and relatively wealthy consumers and city dwellers (who, among other things, are quick to notice the discomforts of urban pollution) despite modest average-national-income levels.

Secondly, a continuous surveillance of fast-changing product requirements and an understanding of the implications of those are essential, not only to short-run trade and marketing decisions, but also to long-run strategic planning for investment.

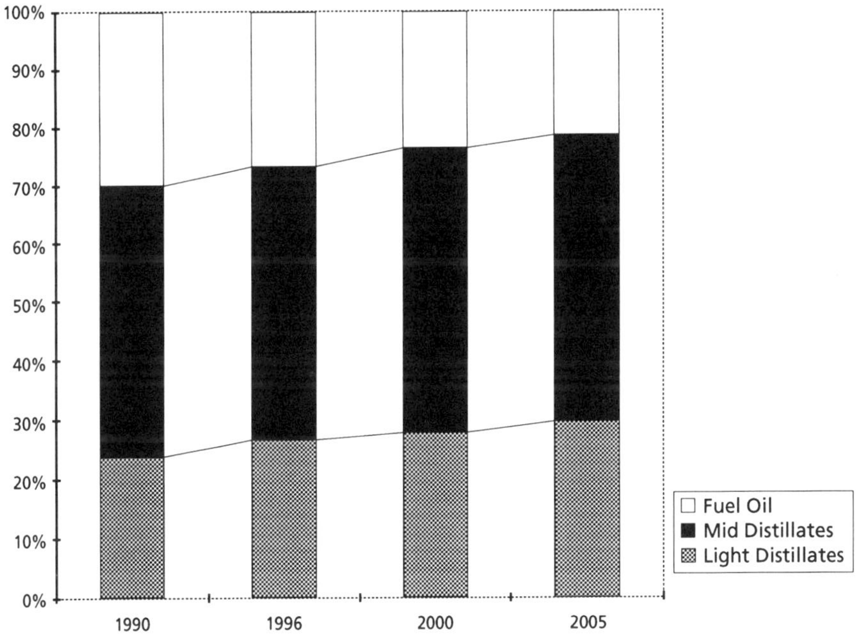

Source: Economics and Business Environment Group, ARCO Corporate Planning

Fig. 4.4 Asia Product Demand (percent distribution)

Both markets and legislation will force the region's refiners to make difficult and inherently risky choices about the scale and type of refinery facilities in response not only to the quantitative demand and supply fundamentals, but also to qualitative environmental constraints. Although the impact of tighter product specifications on refiners in Asia Pacific is far less onerous than that on their counterparts in the US and North-West Europe, substantial changes will be required in three areas: reducing lead content in gasoline, and reducing sulfur content in diesel and residual fuel oil.

Motor Gasoline

Table 4.2 describes gasoline specifications by country, according to the main grades currently sold at the pump. Only Japan and South Korea have fully phased out all leaded gasoline sales. Almost all other countries still sell premium leaded grades at the pump.

Table 4.2 Gasoline Specifications for 10 Countries in Asia[1]

Country	Leaded RON[2]	Lead	Unleaded RON
China	97	0.15	91–93
India	87–93	0.15	na
Indonesia	98	0.15	68
Japan	na	na	91–100
Malaysia	92	0.15	97
Philippines	83–95	0.15	95
Singapore	97	0.15	97
South Korea	na	na	95–100
Taiwan	95	0.075	95
Thailand	87–97	0.15	92–97

Note
1 Main grades sold at pump
2 Research octane number

Source: Industry sources

Table 4.3 gives an idea of the rapid changes in gasoline specification that have taken place among some consuming countries in the region. As can be seen, significant progress has been made in the lowering of average lead levels in the region's gasoline market. Singapore, Malaysia, Thailand, India, and the Philippines all plan to phase out leaded gasoline fully by 2000. Indonesia has plans for phasing out leaded gasoline for Java and outlying cities, which essentially cover most of the consuming centers.

There has occurred a major structural change in the Asian gasoline market over the past two years as trade in premium gasoline has given way to trade in predominantly unleaded gasoline. Until 1994, the most actively traded grade in Singapore was 97 research octane number (RON) grams per liter (g/l) 0.15 grade ("Malaysian grade"), which was the regional product-marker. Spot trade of the Malaysian grade has dwindled, and has been replaced by the 95 RON unleaded grade.

Table 4.3 Gasoline Specification Groupings

1989			
High 0.4-0.5+ g/l	**Low 0.25-0.5 g/l**	**Unleaded**	
China	Hong Kong	Japan	
India	Singapore	South Korea[1]	
Malaysia		Taiwan[1]	
Philippines			
Thailand			
1994			
High 0.25–0.75	**Low 0.08–0.15**	**Unleaded >50% market share**	**Fully Unleaded**
India	Malaysia	Hong Kong	Japan
China	Philippines	Taiwan	South Korea[1]
Indonesia	Thailand		Singapore[1]

Notes
1 Some use of 0.15 g/l grade

Source: Industry sources

In response to the regional trend towards less-leaded and unleaded grades, Shell and Mobil refineries in Singapore have expanded and upgraded their facilities to make cleaner gasoline. Singapore refiners' gasoline specifications are outlined in Table 4.4. Since much of the region does not experience distinct winter seasons, gasoline producers generally do not need to make any changes to Reid vapor pressure (RVP). In North Asia, minor modifications are required for winter, with ten pounds per square inch (psi), and summer, with 9psi. Independent storage facilities have also boosted Singapore's capacity to make different grades required around the region by allowing traders as well as refiners to blend gasoline components.

Table 4.4 Singapore Unleaded Gasoline Specifications

Main traded grades	95 RON	97 RON
Lead content (max g/l)	0.01	0.01
Reid Vapor Pressure (psi)	9–10	9–10
Specific Gravity	0.742	0.750/55

Source: Industry sources

Diesel/Gasoil

The greatest volumetric increase in regional demand for petroleum products is for gasoil, which has a number of different uses, unlike gasoline, which is exclusively for automotive use. Diesel is used as automotive fuel for commercial vehicles as well as by the region's coastal shipping industry. It is also important for agricultural machinery use in India and China, and for substantial power generation in countries such as the Philippines, Vietnam, and Indonesia. Gasoil demand in Asia Pacific, reached some 4.7mb/d in 1994, and has been growing at an annual average of over 6 percent.

Table 4.5 gives current gasoil specifications of both Singapore refiners (which constitute a major source of gasoil exports in the region) as well as the tighter Japanese/South Korean specifications. While the common regional standard (excluding bunkers) is set at 0.5 percent measure of sulfur content by weight (wt S), some important consumers such as India consume 1 percent wt S material. Indeed, the 1 percent gasoil market is primarily constituted by India's monthly buy tenders. Since gasoil standards are much tighter in North-East Asia than in South-East and South Asia, Japanese and Korean shortfalls cannot be covered by Singapore supplies. The deficits are covered by term lifting of straight-run gasoil from Algeria and Italy, as well as smaller volumes from the US, particularly the West Coast (USWC) and Gulf Coast (USGC). Thailand's unique specification for the 90 percent distillation-point can be produced by some Singapore refiners. In general, Asia's large gasoil deficit is filled by Singapore and Middle East exports, which have been increasingly supplemented by arbitrage cargoes from the Mediterranean, Europe, and the US. Gasoil specification limits for different countries in Asia are listed in Table 4.6.

Table 4.5 Asian Gasoil Specifications

	Singapore Output	Japan/South Korea Output
Sulfur (% wt)	0.5	0.2
90% distillation point[1]	370°C	337°C
cetane	48	50
pour point	0 to 9°F	-20 to -5°F

Note
1 Thailand has unique specification of 357° C

Source: Industry sources

Table 4.6 Gasoil Sulfur Specification Groupings

0.2%wt S	0.5%wt S	1%wt S
Japan	Singapore	India
South Korea	Malaysia	Vietnam
Taiwan[1]	Indonesia	Pakistan
	Philippines	Sri Lanka
	China	Taiwan
	Thailand	

Note
1 Allows some use of 1% wt S

Source: Industry sources

Fuel Oil

In the Asia-Pacific region, most of the high-sulfur fuel oil is used as boiler feed for the burning market. The region's limited upgrading capacity, along with large demand for burning fuel, has meant that most incoming cargoes of straight-run fuel oil is used as blendstock rather than refinery feed. For the Singapore market, a major portion of the fuel oil is meant for the bunker market, which applies no sulfur specifications. However, in the residual burning market, stricter sulfur specifications are changing the structure of the market. The regional market for low- and medium-sulfur fuel oil has developed over the past 3–4 years due to the progressive reduction of sulfur content in fuel-oil among many major consuming countries. This development is likely to leave the region short of medium- and low-sulfur fuel oil.

Singapore and Thailand are converting their inland use of high-sulfur fuel oil of 3.5 percent wt S to intermediate grades of approximately 2 percent wt S, and Indonesia, Malaysia, and Thailand are progressively reducing their heavy reliance on high-sulfur fuel oil for power generation by increasing use of natural gas. Even India has announced intentions of reducing sulfur content to 0.25 percent by the year 2000. Fuel-oil specifications by sulfur content are listed in Table 4.7 for eight Asian countries.

Table 4.7 Fuel Oil Specifications

Country	Percentage wt S
China	3.5
Malaysia	3.5
Philippines	3.5
Singapore	2.0–2.8
South Korea	1.6
Taiwan	0.5[1]
Thailand	2.0[2]
Vietnam	3.0

Notes

1 Use of 1% allowed for specific small-volume users

2 In Bangkok area; 3% for elsewhere

Source: Industry sources

Refiners' Responses

Changes in the region's crude output will not have a significant impact on overall crude supply, in that Asia Pacific will remain heavily dependent on Middle East crudes, and this dependence will grow as product demand expands rapidly. Most estimates for Far East low-sulfur-crude production suggest little change in total volumes available over the medium- and long-term, so that most baseload demand will have to be sourced from the Middle East. The crude slate, with larger volumes of refinery runs, will tend to be higher in sulfur (though not necessarily heavier in density, since revenue maximization moves by major OPEC members in the Middle East have increased availability of light and super-light streams). The issue of incremental crude imports, however, may begin to show interesting characteristics as cargoes of North Sea, Alaskan, and West African crudes find their way into East Asia as arbitrage, and diversification opportunities present themselves.

The major impact will be rather on quality issues. The region's crudes are predominantly low in sulfur, and the only alternatives for new supply are West Africa, Libya and the North Sea. Regional sulfur levels in crude average about 0.2 percent wt S, and this is even lower if China's Shengli crude stream is excluded. The regional crude slate's average sulfur levels are expected to remain fairly constant until 2010, according to the East-West Center. Yet at much higher levels of base consumption now, there is a big jump in elemental sulfur volumes.

Table 4.8 Elemental Sulfur Projections in Regional Crude Slate

Year	m tons
1985	5
1994	8
2000	10
2010	16

Source: Fesharaki Associates Consulting and Technical Services, *Refinery Balances and Product Flows: The Asia-Pacific Region*, 1995

The problem of sulfur removal is much larger when the demand side is considered: tightening sulfur specifications on gasoil and fuel oil are gaining momentum in the region. While in the 1970s, most countries allowed 3.5 percent wt S in fuel oil, today many have introduced limits of 2 percent, 1 percent, 0.5 percent and even lower. Indeed, one of the most serious challenges facing refiners in this region is the increase in the sulfur content of refinery feedstock at a time of restrictions on product specifications on the sulfur content of refinery output. Already, the impact of sulfur specifications has affected crude-purchasing patterns among important consuming countries: for instance, Japanese and South Korean refiners have been increasing their off-takes of mid-sulfur grades such as Oman.

Most forecasts of new refinery capacity to come onstream in the region by 2000 project a larger increase in cracking capacity relative to new base distillation, thereby increasing overall upgrading flexibility. Many of the cracking facilities are due onstream later on in the decade, while most of the new distillation units are expected to be up by 1997-98. Furthermore, while there are plans to build new hydrocrackers as well as catalytic crackers, most of the plans for new hydrocracker units remain more tentative. Few units are actually being built. The relative costs and benefits of the two types of technologies have been extensively discussed. For the immediate purposes of this chapter, it is enough to note that most refiners in the region have opted for cheaper cat-cracking configurations that include intensive hydrotreating to improve mid-distillate qualities.

Table 4.9 Asia-Pacific Refinery Capacity and Expansions (mb/d)

	CDU	Coking	FCC/ RCC	HDC	ThC/ Visb.	Cat. Ref.	HDT	VGO Desul.	Resid. Desul.
1995 base	15.67	0.36	2.24	0.53	0.66	1.63	3.16	0.99	0.56
Add. by 1997	2.16	0.03	0.28	0.10	0.06	0.14	0.57	0.11	0.18
Add. by 2000	1.12	0.06	0.40	0.15	0.06	0.08	0.22	0.03	0.15
Total by 2000	18.95	0.45	2.92	0.78	0.78	1.85	3.95	1.13	0.89

Source: Fesharaki Associates Consulting and Technical Services, 1995, *Refinery Balances and Product Flows: The Asia-Pacific Region*

While there is a substantial increase in cracking capacities expected over the second half of the 1990s, there will also be a large increase in distillate hydrotreating capacity, reflecting both the expectation of a more sour crude slate on the supply side as well as progressively tighter restrictions on sulfur content. There is also a substantial increase in residual desulfurization, partly due to the increasingly tighter specifications for inland fuel-oil use as well as for treating feedstocks for residual catalytic-cracking units. In order to phase out lead in gasoline, substantial catalytic reforming capacity will also be installed. While reformate addition will increase benzene levels to the gasoline pool, this is not expected to be an issue outside of the richer countries of North-East Asia (and even here, the issue of benzene content is not clear). Alkylation, isomerization and the use of oxygenates, such as methyl tertiary butyl ether (MTBE) and ethyle tertiary butyl ether (ETBE), will also become increasingly important. Fluid catalytic cracking units, providing refiners with an ability to produce in-house oxygenates, will also gain in importance in Asia's refining configurations.

Yet it is apparent that, in general, Asia-Pacific refiners lack adequate purification processes. South Korea is already exhibiting straight-run surpluses of poor-quality gasoil and fuel oil. As some have noted, it is somewhat ironic that just as the country is switching to low-sulfur specifications, large surpluses of high-sulfur gasoil are emerging, while the country itself has to import low-sulfur supplies from as far away as Algeria and the USWC.[6] The problem is not just restricted to the region's refiners. India's recent moves to source some low-sulfur gasoil have been perceived by some industry observers as a portent of the difficulties that face traditional Middle East suppliers of high-sulfur gasoil that have limited abilities to produce significantly larger volumes of low-sulfur material.[7] Already, the Asia-Pacific gasoil market is two-tiered, according to sulfur content, and this will complicate the gasoil trade patterns in the region.

Of the estimated US$12bn downstream investment by five major oil companies in Asia during 1985–93, the Singapore refineries constituted a third, or US$4bn, of these expenditures. At least another US$1.5bn will have been spent by the Singapore refiners between 1994 and 1996. Over 1993–94, Shell, Mobil, Esso and Singapore Refining Company (SRC) announced a spate of new construction to produce lighter and lower-sulfur products. Thus, Singapore's hydrotreating and desulfurizing capacity will have increased by over 15 percent from early 1994 to end-1996.[8] These investments will ensure that Singapore remains Asia's leading incremental supplier of petroleum products by emphasizing optimal flexibility — on both input and output sides — and the ability to supply niche markets that require higher-quality products. While Singapore is not in a position to increase base capacity, due to the shortage of land — current conditions, in any case, make it questionable whether new base capacity for product exports is economically feasible anywhere in the region — it is in a position of strength to meet new and stringent product specifications emerging in many countries in the region.

The Future of the Asian Oil Industry

It has been estimated that US refiners spent US$20bn over 1990-95 to meet the 1990 Clean Air Act requirements.[9] Despite high capacity-utilization rates, American refiners will almost certainly find it difficult to recover this investment, given low margins and stiff downstream competition. In North-West Europe, the oil industry's main concern is with environmental restrictions, and the consequent costs of improving fuel quality, on the policy horizon in most of the EU countries, and this at a time when there is significant overcapacity for gasoline production. What can refiners expect in Asia Pacific?

Clearly, aggressive reformulation of motor gasoline along the lines of the US — mandatory oxygenation, and benzene and aromatics regulations — are not expected in most Asia Pacific countries. Yet the medium-term outlook for refinery margins remains poor, at least until the current spate of distillation-capacity building is absorbed by Asian product demand growth in 1998-99. While the region's refiners face major challenges with respect to rapidly changing product specifications, it is important to note that this is just one aspect of the difficult new environment that has emerged since 1993.

Refiners in Asia Pacific need to attenuate the imbalance between the structure of demand and supply for petroleum products; in particular, they need to address the large deficit in the middle of the demand barrel, especially that of gasoil. They need to achieve this in the context of increasing dependence on Middle East crude as low-sulfur Asian crudes become increasingly constrained relative to growing product demand. In short, refiners face both quantitative (specifically in gasoil) and qualitative requirements. And, most difficult from the project-financing point of view, the refining industry faces these challenges in an environment where returns on refining assets are universally poor.

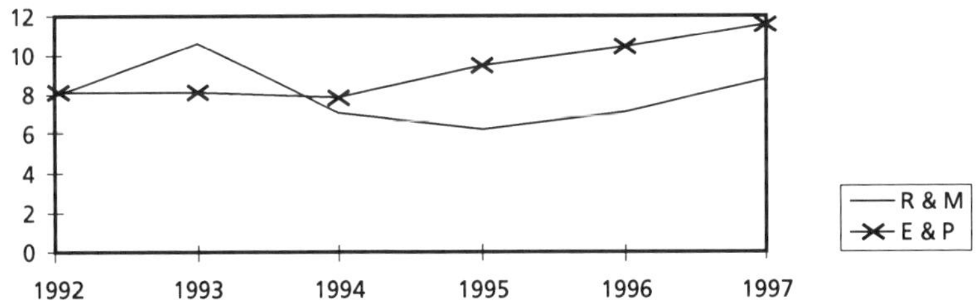

Source: NatWest Securities, 1997

Fig. 4.5 Return on Fixed Assets (percentage) – estimate and forecast for percent return on fixed assets for exploration, production and refining, and marketing for 15 of the largest private oil companies.

As the demand slate changes in the region, refiners will need to construct not only distillation capacity, but also, more importantly, a substantial portion of new conversion capacity. One could expect long-residue catalytic crackers, hydrocrackers, and hydrodesulfurizers as the "core configuration" of rejuvenated refineries, adding substantially to unit refining costs.[10] Deep hydrodesulfurization will become more important, and the increasingly sour crude slate, combined with larger volumes of domestic refinery runs, will force residual desulfurizing on a large scale, further adding to refinery costs. Countries, and refiners, will have little choice over much of this requirement. In a low-refining-margin environment, the financing of new equipment, and especially grass-roots facilities, will pose major challenges to both governments and the private sector.

Investment requirements for basic process equipment alone in the region's refining sector have been estimated at US$50bn dollars in the decade 1995–2005. This is a conservative estimate, and the more likely figure — once requirements for infrastructure, complex configuration and clean fuels are considered — will be closer to US$75–US$100bn in today's dollars.[11] Financing these large capital requirements will be a major task for governments in the region, if not beyond their capabilities altogether. The Asian countries requiring the largest investments — particularly its largest consumers China, India, and Indonesia — will find it even more difficult to raise such sums without undergoing a fundamental policy change (or more appropriately, a reversal) toward: domestic crude and product pricing; foreign investments in the upstream and downstream petroleum and infrastructure sectors; and institutional rules of competition and commercial contract.

For countries with higher income levels, and greater proportions of middle and professional classes among their urban populations, the challenge for governments will be to enact product-specification standards that are optimal. In some cases, it may be more efficient to require better engine maintenance, or to desulfurize flue gases off stacks, rather than enact stringent clean-product regulations.

As customers of crude oil, refiners and their capital investment programs will play a major role in determining the demand for different grades of crude oil. The new capacity coming onstream in Asia in particular is already having an impact on international crude oil trade flows. Changes in product specifications will impose new requirements, not only on refiners, but also on producers of crude oil. For instance, the relatively simple configuration of the refining industry East-of-Suez has led to a large increase in light West African crude flows eastwards, a development that would have been unexpected just a few years ago. Smaller movements of North Sea and Alaskan crudes into East Asia have also been observed recently.

Refiners will be forced by both legislation and the market to invest in upgrading facilities to match demand patterns and required product specifications. Only the efficient refiners will be better able to absorb the costs of progressively tighter environmental legislation. The likely response of the oil industry will be to pursue joint ventures with NOCs and host governments, to form partnerships with local firms well versed in the local business culture of different countries and regions, to form mergers, consolidations, and "strategic alliances" (viz. Mobil and BP in European downstream operations, or the Shell-Texaco-Saudi Refining merger in the US) for cost reductions, and perhaps to attempt further vertical integration in the quest for optimal portfolios across the petroleum chain. With the growing multiplicity of actors in the oil industry, the era of the giant, integrated oil companies unilaterally determining the shape of the industry has passed.

CHAPTER 5

"Back-door" Privatization Initiatives in the Gulf: An Inadequate Step in the Right Direction

Wael Al-Mazeedi

Overview: "Back-door" vs "Front-Door" Privatization

The long-term health of Gulf economies will depend on governments' willingness to target core economic areas, such as the energy sector, for deregulation, commercialization, and privatization. So far, governments have been reluctant to open their energy industries fully to private ownership and management. The consequent inadequate rate of re-investment in hydrocarbon exploration and development — as compared with other private oil companies of a comparable size — has now been recognized and is being corrected.

Although national oil companies (NOCs) continue to exercise their monopolies over core energy-industry activities, a substantial number of upstream and downstream grass-roots project opportunities have become available for domestic and foreign private-sector participation. These opportunities range from the farming out of specific oil and gas production-related activities and the privatization of local retail-petroleum distribution companies, to the setting up of joint ventures with foreign partners to undertake energy investment projects both within the Gulf and abroad. These peripheral privatization initiatives constitute "back-door" privatization.

While these "back-door" privatization initiatives represent steps in the right direction, they fail to maximize the potential efficiencies, managerial improvements, technological innovations, and overall economic dynamism that full competition would promote in — indeed, impose upon — the energy sector. Moreover, "back-door" privatization restricts the development of a healthy symbiosis between the energy and financial sectors, among others.

The Gulf has a large financial sector, but that sector has not focused on infrastructure investment or project finance, tools that are fundamental to the energy industry worldwide. The Gulf financial sector is developing these tools — specific examples will be cited later in the chapter — but such development would be accelerated in the event of full privatization of the NOCs.

For this reason, "back-door" privatization, although it does allow for the development of a small upstream and downstream private petroleum industry, effectively cheats the local economy twice: by failing to take advantage of the incentives that a market economy would impose on the core hydrocarbon sector itself, and by failing to take advantage of the potential symbioses between the two most important economic activities in the Gulf, namely the oil industry and the financial sector. "Back-door" privatization alone fails to provide the basis for a flexible, dynamic, interdependent private economy on which the Gulf, in the long-run, must rely. Numerous additional symbioses, similar to that between energy and finance, would be stimulated by full "front-door" privatization, enabling the energy sector to act as a springboard for long-term, self-generating economic growth in the Gulf, growth that must continue well beyond the hydrocarbon era.

As highlighted in a previous article,[1] Gulf NOCs at the beginning of the 1990s already lagged well behind the rest of the world's movement toward privatization. The lag is even greater today, for only "back-door" privatization has been pursued in the Gulf,[2] while NOCs elsewhere have been fully privatized. The earlier article described the potential direct benefits of full privatization of Gulf NOCs, including enhanced efficiencies, improved technical and managerial expertise, and replenished government coffers. This chapter focuses instead on the broader, more indirect, and longer-term benefits that can be achieved only through full privatization.

Table 5.1 National Oil Companies in the Gulf: A Comparative Assessment of Vital Statistics

	Company Name				
	KPC	ADNOC	PDO	QGPC	S Aramco
Ownership Status	State-owned	State-owned	60% State-owned	State-owned	State-owned
Type of Firm	Integrated	Mainly upstream	Mainly upstream	Mainly upstream	Integrated
Diversification	Chemicals	Chemicals	Chemicals	Chemicals	Chemicals
Revenues[1] (US$m)	14,000	N/A	N/A	N/A	N/A
Employees	11,000	85,000	3,000	55,000	43,000
Integration Ratio[2]	221.3%	622.4%	586.3%	736.5%	487.3%
Oil Reserves (m/b)	96,500	64,452	2,820	3,376	261,374
Gas Reserves (b/cf)	52,965	137,102	12,464	317,260	190,215
Production Output (tb/d)	2,907	1,276	469	464	8,574
Downstream Ratio[3]	74.9%	125.4%	75%	187.3%	122.3%
Refining Capacity[4] (tb/d)	948	205	80	63	1,760
FCC Conversion Ratio[5]	219%	109%	None	None	74%
Retail Outlets	5,377	None	None	None	Joint Venture

Notes

1 With the exception of KPC, other Gulf NOCs do not publish financial statements
2 Liquid production divided by refining capacity
3 Total product sales divided by refining capacity
4 Worldwide
5 Equivalent fluidized-bed catalytic cracker-conversion capacity of the refinery, divided by total atmospheric distillation capacity

Source: *Ranking The World's Top Oil Companies*, 1996 edition, by Booz Allen & Hamilton and *PIW*, and based on 1994 data

From Comparative Advantage to Competitive Advantage

Gulf countries still hold a comparative advantage, even an absolute advantage, in hydrocarbon production.[3] This advantage has proven extremely valuable. For twenty years, Gulf governments tapped the cash-flows of NOCs in order to subsidize other sectors, and especially to invest in modern public infrastructure: highways, telecommunications, housing, education, medical care, and so forth. The consequent inadequate re-investment in hydrocarbon exploration and development has now been recognized, and is being corrected.

The Gulf's hydrocarbon advantage entails certain hazards. Firstly, it continues to cushion NOCs from the rigors of market forces. Secondly, it focuses the attention of both public and (to the extent they are admitted) private investors in the Gulf on the short-term profits to be earned from resource-based investments, and reduces the attention paid to longer-term regional economic health.

The present resource-based comparative advantage of the Gulf represents an opportunity to turn comparative advantage into a longer-term competitive advantage.[4] Lower feedstock costs will not forever offset higher unit costs for plant construction, operation, and marketing. Gulf NOCs are vulnerable to factor cost-swings, such as crude oil prices and the exchange rate of the US dollar, the currency in which oil is denominated, as well as protectionist policies in developed consumer markets.

Some Gulf NOCs are trying to convert their resource advantage into a durable competitive advantage through vertical integration outside the Gulf.[5] How well this strategy will succeed remains to be seen. The costs of keeping up with industry-wide technological change and increasingly stringent environmental standards may be difficult without continuous subsidization from the NOCs' upstream resource advantage. Some NOCs paid premium prices for assets that the international majors or independent oil companies no longer found viable. While benefits of this integration process have been realized (particularly in expanding the Gulf NOCs' knowledge of non-Gulf markets, and in reducing the political risks of selective-market closure), the costs of bringing newly integrated systems up to competitive standards has been high. Furthermore, Gulf NOCs refrained from jointly pursuing downstream opportunities with other Gulf NOCs, thereby depriving themselves of synergistic opportunities.

Many unexploited opportunities exist for Gulf NOCs along the downstream value-chain (Fig. 5.1). Besides the obvious refining and petrochemical — thermoplastics and synthetic fibers — there are those targeted at more differentiated and sophisticated markets, such as fine and specialty chemicals. Furthermore, since gas is more closely tied to infrastructure, a host of immediate opportunities emerged in those countries with a large appetite for energy consumption. Examples include

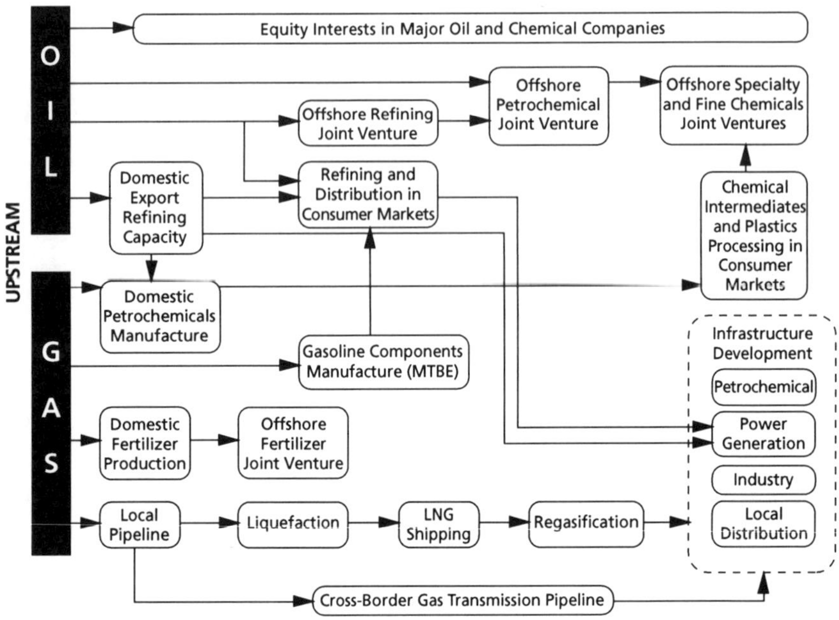

Fig. 5.1 Oil and Gas Downstream Value Chain

integrated energy projects involving gas transmission and distribution, LNG regasification, and power-generation, among others.

Private management would have moved faster to seize these opportunities than any of the NOCs have done. The NOCs will point proudly to their downstream assets and joint ventures, but compare that, for example, with how rapidly the European, and especially US, refining-marketing industries have restructured themselves in any particular twelve month period; similarly with the way in which Japan's refining industry has responded to the liberalization of Japanese petroleum products imports. Market forces require private management to anticipate changes and to react very rapidly. In comparison, the moves of Gulf NOCs seem ponderously slow.

Downstream integration alone, however successful it may turn out to be in a commercial sense, will not be sufficient to convert the Gulf's present comparative advantage into a long-term competitive advantage. By itself, it will not bring to the Gulf the skills, and especially the incentives, required for the emergence of a dynamic and self-generating post-hydrocarbon economy.

A second policy by which some Gulf countries have sought to convert comparative advantage into sustainable competitive advantage has consisted of promoting "economic diversification." Substantial public-sector resources are currently channeled toward the creation and subsidization of entirely new commercial activities in the name of "diversification". That Gulf economies should become more diversified is undoubtedly desirable, but whether this can best be achieved — or achieved at all — through government targeting and subsidization of individual new industries is subject to question. Would it not be easier, more natural, more efficient, and more sustainable, were economic diversification to emerge from the complementarities and symbioses surrounding a privatized petroleum industry?

As long as the petroleum sector remains government-owned, it remains isolated. Local equipment purchases or the subcontracting of selected services, sometimes undertaken by NOCs to encourage the development of local petroleum-service companies, only scratches the surface of potential complementarities.[6] NOCs too rarely look outside themselves for investment banking, insurance, legal services, and high-level consultancy or management services. They infrequently interact with academic researchers, losing thereby the benefits of mutual stimulation. Perhaps most significantly, NOCs fail to attract and nurture the best entrepreneurial talent, as individuals with entrepreneurial instincts prefer to seek their fortunes outside the bureaucratic NOC structure, and sometimes outside the Gulf entirely. NOCs even find themselves isolated from other branches of their own government, for regulatory or policy reasons.[7]

Nurturing Natural Symbioses

Consider, for a moment, the cities of Houston and Singapore. Both are home to important oil-related industries: refining, storage, transportation, and petrochemical manufacturing. But both cities also have become economically diversified. In particular, both are financial centers, both are involved heavily in information technology and telecommunications, and both cities also have other significant commercial and academic competencies. Moreover, the diversified economic activities in each city are interdependent. The different economic sectors grew in response to, and were stimulated by, each others' needs and offerings.

Even a cursory look at Houston and Singapore raises the question of why the Gulf has not also evolved a more diversified and interdependent economy, one which is less sensitive to hydrocarbon price shocks. It also raises the question of what it would take for the Gulf to achieve such an evolution. The answers are clear: nationalization of the petroleum industry has inhibited the natural evolution, and privatization would stimulate it.

Evidence that the private sector is alive and well in the Gulf is abundant. Moreover, the "back-door" privatization steps, however hesitant and partial they may be, reveal the eagerness and ability of the private sector to enter the hydrocarbon industry and to exploit the natural complementarities between energy and finance. An appendix to this chapter includes some of the more notable examples of private energy-finance cooperation.

Consider the Equate project in Kuwait, and the Ras Gas project in Qatar, both summarized in the appendix. These projects represent the first wave of limited-recourse — off-balance-sheet — multisource project financing transactions in the hydrocarbon sector in the Gulf [8] (Figs 5.2 and 5.3). This type of financing differs from the refining and petrochemical joint ventures at Jubail and Yanbu in Saudi Arabia, which were financed through a combination of shareholders' equity and corporate loans. As indicated in the appendix, domestic Gulf-based financial institutions have participated in lending to the Equate and Ras Gas projects.

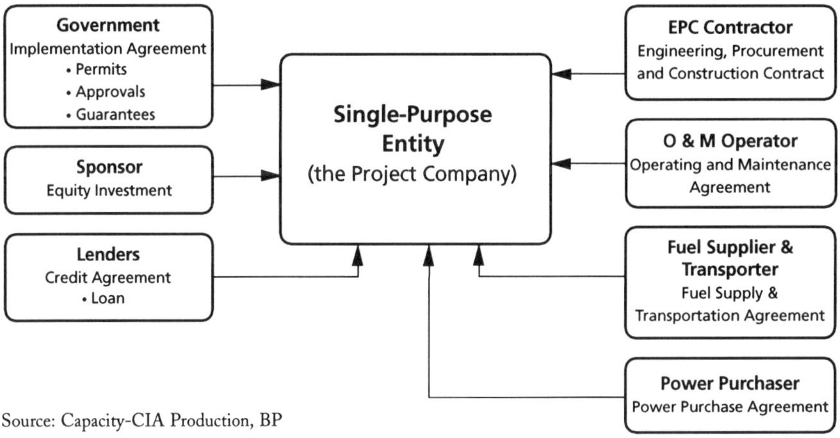

Source: Capacity-CIA Production, BP

Fig. 5.2 Limited Recourse Project Finance-Contractual Arrangements — "Typical Independent Power Project"

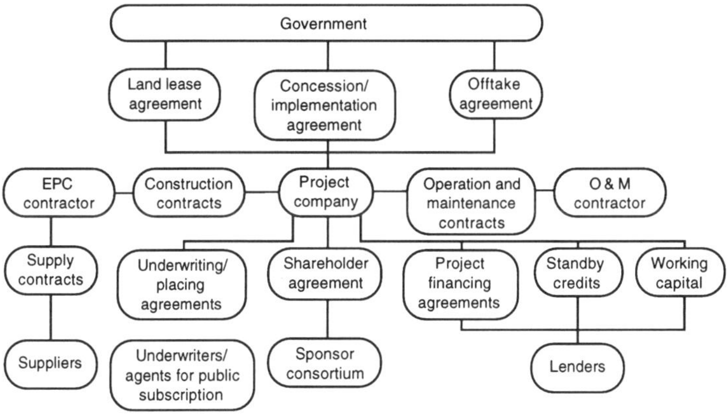

Source: *International Financial Law Review*, November 1995

Fig. 5.3 Limited Recourse Project Finance Generic Contractual Arrangements

Equate and Ras Gas in particular set a precedent representing a benchmark for future project-financing in the region. Equate's financing package linked a syndicated loan with an Islamic lease under one inter-creditor agreement. Ras Gas included an international bond offering alongside its commercial bank and export-credit-agency financing. The bond offering for Ras Gas is the largest project-finance bond ever issued in the international capital markets. Also, it is the first time that a Gulf project has been financed in the capital markets, and the first rated, long-term bond from the Middle East.

Consider also the examples of Asia Petroleum in Pakistan and Preem Petroleum in Sweden, cited in the appendix. As these examples show, Gulf-based private investors are pursuing downstream opportunities abroad with the intention of integrating backwards, building secure and profitable disposal outlets in consumer markets for Gulf crudes and refined products.

Nor are private Gulf-based companies limiting themselves to downstream petroleum-related investments abroad. As the cases of Xenel Industries and Delta Oil Company show (see appendix), Gulf-based entrepreneurs are developing and investing in a wide range of energy transformation and transportation projects outside the Gulf. Although these projects are not necessarily linked to hydrocarbon reserves in the Gulf, they indicate a strong desire by the domestic private sector to be engaged in the energy business as a whole, and more specifically, a commitment toward building a track record in significant international energy-related project undertakings. It would be a shame not to encourage such entities to pursue energy-related activities in their own backyard. All of these examples suggest that an alert and eager private sector would readily invest in the core hydrocarbon production industries of the Gulf, were these made available for private investment.

To those who suggest that full, "front-door" privatization of companies as large as some of the Gulf NOCs is not feasible because of their size, one can only suggest that the example of Petronas Gas in Malaysia be studied with care.

Conclusion: Beyond Energy-Finance Symbioses

In summary, Gulf governments should build on the region's current comparative advantage in the hydrocarbon sector. The competitive positioning of the hydrocarbon industry in the Gulf can only be strengthened if the private sector is allowed to participate actively in all its facets. Through privatization, governments will nurture horizontal linkages that emerge naturally between the hydrocarbon sector and other sectors of the national economy, such as banking, financial services (including Islamic banking), insurance, legal services, academia, consultancy, training, and so forth. "Back-door" privatization today only scratches the surface of such opportunities.

The long-run vitality of the Gulf will depend upon whether the hydrocarbon industry is allowed to become the foundation of a self-sustaining and self-generating Gulf business center. Studies of current centers of business activity such as Silicon Valley in California have emphasized the importance of cross-firm and cross-sectoral exchanges, of innovative ideas, of problem-solving approaches, and of mutual services.[9] Regions lacking such cross-sectoral reinforcements have clearly lagged.

Gulf planners and researchers seeking to promote continued economic vitality beyond the hydrocarbon era must identify the symbioses that can be encouraged, and the government postures that will contribute to this goal. Privatization, relaxation of investment restrictions, and a general reduction of government involvement in the economy will prove to be essential.

Appendix
Examples of Private-Sector Symbioses: Energy Industry – Financial Industry

The following examples illustrate major themes raised in the chapter. These examples demonstrate the enormous potential for:

- full, "front-door" privatization of national energy companies;

- the creation of joint ventures between local energy companies and major international firms;
- overseas downstream integration by Gulf-based energy companies; and, most importantly,
- creative financial interaction between Gulf-based energy companies and Gulf-based financial institutions.

PETRONAS GAS: A SUCCESSFUL "FRONT-DOOR" PRIVATIZATION

Petronas Gas, a Malaysian government-owned entity, was effectively privatized in July 1995. The initial public offering of Petronas Gas on the Kuala Lumpur Stock Exchange was over-subscribed 12 times. Four hundred and fifty million shares and 45 million warrants were offered to institutional investors and the public, of which 135 million were allocated for an international tranche. Investors were 43 percent Asian, 21 percent Malaysian, 27 percent European, 5 percent American and Canadian, and 4 percent from the rest of the world.

Petronas Gas
Amount: Malaysian M$2.4bn
Type of issue: IPO
Number of shares: 450m and 45m warrants
Offer price: M$5.30 (domestic), M$8.00 (foreign), M$3.50 (warrants based on a strike price of M$5.30)
Pricing date: July 31, 1995
Trade date: September 4, 1995
Listing: Kuala Lumpur Stock Exchange
Commissions: 1%
Global lead manager: Rashid Hussain
Global coordinators: BZW, Daiwa, Morgan Stanley

RAS LAFFAN ("RAS GAS") LNG PROJECT: AMBITIOUS JOINT VENTURE

Project Description

Ras Laffan Liquefied Natural Gas Company Limited (Ras Gas) is a Qatari joint stock company formed to produce and sell up to 10 million tons per year (t/y) of liquefied natural gas (LNG) condensate and other hydrocarbon products as well as to develop, finance, construct, operate and maintain a two-train plant.

Unique Features

The US$1.2bn bond issued for Ras Laffan is the largest project finance bond ever issued in the capital markets. Moreover, it is the first time that a Gulf project has been financed in the capital markets. It is also the first rated, long-term bond from the Middle East.

Ras Gas Liquefied Natural Gas Project

Sponsors: Qatar General Petroleum Company (QGPC) (66.5%); Mobil QM (26.5%); Nissho Iwai (4%); Itochu (3%)

Total project cost: US$3.4bn

Tranche 1: US$400m, 10-year bond priced at 135bp over Treasuries

Tranche 2: US$800m, 17-year bond priced at 187.5bp over Treasuries

Tranche 3: US$450m uncovered commercial bank loan with an average life of 9.2 years priced at LIBOR+95bp for years 1–5, LIBOR+175bp for years 6–7, LIBOR+187.5bp for years 8–10 and LIBOR+200bp for years 11–12.75

Tranche 4: US$900m, ECA-backed loans with an average life of 8.7 years priced at standard consensus rates

ECA commitments: US EXIM, US$465m; the UK Export Credit Guarantee Department (ECGD), US$250m; Sace, US$185m

Financial adviser to Ras Laffan: Goldman Sachs

Lead arrangers: IBJ, Credit Suisse

Co-arrangers: Arab Banking Corporation, Bank of Nova Scotia, Indosuez, BNP, Paribas, BZW, Bayerische Landesbank, Cho Hung, CIBC Wood-Gundy, DKB, Fuji, Gulf International, ING, Korea First, RBS, Sumitomo, WestLB

Legal advisors: Skadden, Arps, Slate Meagher & Flom; Latham & Watkins

Source: *Project and Trade Finance* (February 1997) and Bill Voge, partner, Latham & Watkins Project Finance International (January 29, 1997)

EQUATE: A SUCCESSFUL JOINT VENTURE BETWEEN DOMESTIC AND INTERNATIONAL FIRMS

Project Description

Equate is a joint-stock Kuwaiti company formed to develop, finance, construct, operate and maintain a 650,000t/y ethane cracker, a 450,000t/y Unipol process polyethylene plant, and a 350,000t/y ethylene glycol production plant. The project will act as an initial building block for the development of secondary and tertiary petrochemical production involving the private sector. In conjunction with the Industrial Bank of Kuwait (IBK), the Kuwaiti government has identified a list of joint-venture opportunities for private sector investment. The government, through the Petrochemical Industries Company (PIC), would retain a minority stake in the new private-sector-controlled ventures.

Unique Features

This transaction signals the first serious attempt by the government to encourage private-sector participation in the hydrocarbon sector, albeit only in greenfield projects. This is also the first limited-recourse project-financing in Kuwait, and the first project company of its kind to offer 10 percent of its outstanding shares at financial close to the public by way of an IPO on the Kuwait Stock Exchange.

Equate Petrochemical

Sponsors: PIC, a subsidiary of Kuwait Petroleum Corporation (45%); Union Carbide Corporation (UCC) of the US (45%); and Bubiyan Petrochemical Corporation, a publicly traded company (10%). PIC,

UCC, and Bubiyan are providing US$730m in equity and subordinated debt.

Debt raised: US$1.35bn

Arrangers: Kuwait Finance House, Arab Banking Corporation, Apicorp, Bayerische Vereinsbank, Chase, CIBC, Citibank, Gulf International, Gulf Investment Corporation, Industrial Bank of Kuwait, JP Morgan, Kuwait Real Estate Bank, National Bank of Kuwait, Sumitomo, Wafabank.

Underwriters: NBK, GIC

Term loan: US$500m (US$200m Islamic lease arranged by Kuwait Finance House)

Tenor: 10 years, 7 months

Repayment: 17 semi-annual

Term loan: US$700m

Tenor: 8 years, 7 months

Repayment: 13 semi-annual

Margin for both tranches: LIBOR+175bp if project completed by September 30, 1998; 187.5bp thereafter

Commitment fee: 50bp

Bridge facility: US$150m

Repayment: On completion

STAR ENERGY REFINING COMPANY LTD: AN OFFSHORE, DOWNSTREAM VENTURE

Star Energy is a recently commissioned 130tb/d deep-conversion refinery located in the Rayong Province (Thailand). The refinery is designed to maximize production of unleaded gasoline and low-sulfur diesel fuel for the domestic Thai market. The cost of the project is US$1.85bn. The Oman Oil Company (OOC), which subsequently withdrew as a shareholder, had an aggregate equity interest of 25.6 percent. The remaining shareholders include Caltex Trading and Transport (64 percent) and the Petroleum Authority of Thailand (36 percent). OOC had an exclusive crude-supply agreement with the project company.[10]

Star Refinery Financing Plan

	US$m	%
Equity		
CXO (Caltex/Oman Oil Company)	355.2	19.0
Petroleum Authority of Thailand	199.8	11.0
Total Equity	555.0	30.0
Debt		
IFC "A" Loan	75.0	4.0
IFC "B" Loan	275.0	15.0
Parallel Commercial Loan/Guarantee	449.0	24.0
Export Credits	400.0	22.0
Thai Revolving Credit	96.0	5.0
Total Debt	**1,295.0**	**70.0**
Total Project Cost	**1,850.0**	**100.0**

AL-MANAH POWER PLANT: PRIVATIZING THE ELECTRIC POWER SECTOR

Al-Manah is a 90 MW independent power project (IPP) including 200km transmission lines in Oman, which was awarded by the government on a build-own-operate-transfer (BOOT) basis, and which reached financial close in 1995.[11] Sponsors include Tractabel of Belgium (12 percent), its investment arm Powerfin (20 percent), Omani companies (22 percent), and the International Finance Corporation (IFC) (3 percent).

An Omani joint-stock company, the United Power Company was established to implement the project. The financing was arranged by Arab Bank and Paribas. Coface and ECGD were also involved. The project is supported by a 20-year offtake contract with the Omani Ministry of Electricity and Water. The government remains in control of transmission and distribution of power to consumers. The total cost of the project is US$213m, of which US$69m is a loan from the IFC, US$72.0m is lent by export credit agencies, and US$72.0m is equity: 40% of which has been raised via an IPO on the Muscat Securities Market. This is the first example of multilateral participation — through the IFC — in the Gulf. The bank debt is raised via the IFC "A" and "B" loan structure.

ASIA PETROLEUM LTD: PRIVATE INFRASTRUCTURE INVESTMENT ABROAD BY A PRIVATELY OWNED GULF-BASED COMPANY

Asia Petroleum Ltd was established in 1994 to develop, design, finance, construct, own, and operate pipelines linking oil-storage facilities to major oil-fired power-generating centers in Pakistan. Its first project is to finance, construct, operate, and maintain an 80km pipeline to transport fuel oil from an import-receiving terminal at Port Qasim to a 1280MW independent power project at Khalifa Point in Baluchestan: the Hub Power Project. The total project cost is US$100m, and project sponsors include private-sector entities from Singapore (Promet), the US (Veco International), Kuwait (Independent Petroleum Group [IPG]), and Pakistan (Pakistan State Oil [PSO]). The World Bank is providing a US$30m subordinated debt facility through the Pakistan Private Sector Development Fund, which is managed by Pakistan's private sector on-lending arm, the National Development Finance Corporation. The project is likely to play a major role in alleviating supply-logistic constraints and developing a modern and efficient petroleum-products transportation infrastructure in Pakistan. The IPG was responsible for the fuel-oil supply for the power plant, and for this purpose secured a long-term supply contract with a Gulf NOC for the duration of the project's concession. In order for IPG to continue pursuing overseas infrastructure opportunities backed by fuel-supply contracts it needed access to additional capital. In 1995, it successfully raised capital through an IPO on the Kuwait Stock Exchange.

PREEM PETROLEUM AB: A MAJOR INTERNATIONAL RETAIL AND REFINING ACQUISITION BY A PRIVATELY OWNED GULF-BASED FIRM

In 1994, the Al-Amoudi Group of Saudi Arabia, through Corral Petroleum Holdings, acquired OK Petroleum in Sweden for US$1.2bn. OK Petroleum, the largest oil company in Sweden, was owned by the Swedish government, the Finnish state-owned oil company Neste Oy, and the Swedish Cooperative Union KF. Following the acquisition, OK Petroleum secured a long-term crude-supply contract with Saudi

Aramco, and was renamed Preem Petroleum AB. Preem Petroleum operates two refineries with a combined capacity of 265tb/d and a retail network of 1,028 service stations, giving it 25 percent of the Swedish retail market. The company had revenues of US$2.6bn in 1995 and employs 4,200 people. Nimir Petroleum, another private sector Saudi Arabian company, has been awarded the management contract to run Preem Petroleum.

The experiences of Asia Petroleum Limited and Preem Petroleum demonstrate how the private sector was able to develop, finance, and operate offshore downstream opportunities. In essence, the private sector is complementing Gulf NOC efforts by securing disposal outlets for their crude and petroleum products through the acquisition of offshore downstream assets. In return, the private sector is offered access to long-term supplies of crude and/or products on commercial terms.

Xenel Industries & Delta Oil: Private investment abroad by Gulf-based firms

Other private-sector entities expected to play a role in downstream development are Xenel Industries and the Delta Oil Company, both of Saudi Arabia, along with Crescent Petroleum and Star Energy of the United Arab Emirates. Xenel has been a lead developer in the US$2.0bn Hub Power Project in Pakistan, considered the first large multisourced limited-recourse project financing in the developing world. The project is now fully funded. In a joint venture with Neste of Finland, Xenel is also involved in the development of an 800,000t/y methyl tetra butyl ether (MTBE) plant in Saudi Arabia. Delta, with Unocal Corporation of the US, is developing the US$2.5bn (1,300km, 2bn cubic feet per day) Turkmenistan–Afghanistan–Pakistan gas pipeline project. Crescent, with Brown & Root, Trans-Canada Pipelines, and Itochu Corporation, as the Gulf-South Asia Corporation, is developing a US$3.5bn (1,700km, 1.6-2.4bn cubic feet per day) Qatar–Pakistan gas pipeline.

Contributors

WAEL AL-MAZEEDI is a consultant specializing in oil and gas privatization and project financing who has worked on a variety of assignments in the Middle East and South Asia regions. He is co-founder of the MENA Task Force for Private Infrastructure Finance, and a former consultant to the World Bank. Mr Al-Mazeedi started his career with the Major Projects Group of Kuwait Petroleum Corporation's Corporate Planning Department as an investment analyst, where he reported on the corporation's energy and fertilizer-related foreign direct investment in Europe, North Africa, Turkey, and the People's Republic of China. He left the public service sector in 1990, and established the Strategic Investment Forum, a London-based management consultancy firm engaged in privatization and professional management development. Mr Al-Mazeedi holds a BA in Applied Chemistry and Control Engineering from the University of Kent, and a Post-graduate Diploma in Petroleum Economics from the College of Petroleum and Energy Studies in Oxford.

MAJID A. AL-MONEEF is Associate Professor of Economics at King Saud University in Riyadh. He is an advisor in the Ministry of Petroleum and Mineral Resources of Saudi Arabia, and heads the Energy Team of the Gulf Co-operation Council. Dr Al-Moneef also serves as the National Representative for Saudi Arabia on OPEC's Economics Commission Board.

TILAK DOSHI is Director of Economic and Industry Analysis for Corporate Planning at ARCO in Los Angeles. Prior to this appointment he was Head of Research at Louis Dreyfus Energy Asia Pacific Pte Ltd in Singapore and Coordinator of Energy Projects at the Institute of

Southeast Asian Studies. Dr Doshi holds an MA in Economics from Victoria University of Wellington, New Zealand, and a PhD in Economics from the East-West Center, University of Hawaii at Honolulu. Dr Doshi is the author of numerous books and reports, which include *Houston of Asia: The Singapore Petroleum Industry* and *Vietnam: Asia's Newest Oil Exporter.* He is also a regular contributor to the *Far Eastern Economic Review.* Dr Doshi is a founding member of the International Association of Energy Economics in Singapore.

GIACOMO LUCIANI has been the Deputy Director of International Affairs at ENI since 1971. He was educated at the University of Rome, Yale, and Harvard. Before joining ENI, he worked for the Bank of Italy and for the Italian Institute of International Affairs. He has taught at UCLA, and is currently visiting professor at the Institut d'Etudes Politiques in Paris and at the Bologna Center of SAIS, John Hopkins University. He has consulted for various international organizations and Gulf countries. He is the author of several books and articles, including: *The Oil Companies and the Arab World* and *The Arab State.*

MARIO SALUSTRI is a graduate of Rome State University, where he received a BA in Statistical Studies. After a period as a junior researcher at the Institute for Economic Development and Technical Progress (ISVET), he joined ENI in 1970. He is currently senior economist in ENI's International Affairs Department.

PAUL STEVENS was educated as an economist and as a specialist in the Middle East at the University of Cambridge, and the School of Oriental and African Studies. From 1973 to 1979, he taught at the American University of Beirut, interspersed with two years as an oil consultant. In 1979, he began teaching at the University of Surrey, where he was a founding member of the Surrey Energy Economics Centre and joint creator of the Third World Energy Policy Studies Group. Since 1993, he has been BP Professor of Petroleum and Economics at the Centre for Petroleum and Mineral Law and Policy, University of Dundee, Scotland. Professor

Stevens has published widely on energy and development economics and has worked as a consultant for many companies and governments.

KEIICHI YOKOBORI is the Director of the Institute of Energy Economics, Japan (IEEJ) and President of the Asia Pacific Energy Research Center (APERC). Before holding this post, Mr Yokobori was Senior Advisor to IEEJ, as well as Executive Director of the Organizing Committee of the Sixteenth (Tokyo) Congress of the World Energy Council. From 1963 to 1990, Mr Yokobori worked in Japan's Ministry of International Trade and Industry, where he was Director of the Office of Information and Energy Systems Operations. Mr Yokobori was also Chairman of IEA's Committee on Research and Development (CRD), which is now the Committee of Energy Research and Technology (CERT).

Notes on the Text

Introduction: Strategic Positioning in the Oil Industry: Trends and Options

1 See M.A. Adelman, *The World Petroleum Market* (Baltimore MD, John Hopkins University Press, 1972); J. Blair, *The Control of Oil* (London, Macmillan, 1976); J.E. Hartshorn, *Oil Trade: Politics and Prospects* (Cambridge, Cambridge University Press, 1993); D. Yergin, *The Prize* (New York, Simon and Schuster, 1991).

2 This is because OPEC has effectively acted as the residual supplier to the market. All other producers have tended to produce at sustainable capacity.

3 See P. Stevens, "A Survey of Structural Change in the International Oil Industry 1945-1984" in *The Changing Structure of the World Oil Industry*, edited by D. Hawdon (London, Croom Helm, 1984); P. Stevens, "The Determination of Oil Prices 1945-95," *Energy Policy* vol. 23, no. 10 (1995).

4 See J.E. Hartshorn, "From Multinational to National Oil: The Structural Change," *Journal of Energy and Development* vol. 6 (Spring 1980); R. Seal, "Downstream Oil: Deintegrated Profitability or Ruin?" speech reprint (London, British Petroleum, 1991).

5 The capacity is given in calendar days. Hence, the apparent excess capacity in the 1960s simply reflected routine downtime to allow for maintenance.

6 See K. Hamm, "Downstream Business: Economics and Future Prospects," Middle East Oil & Gas: Towards the Third Millennium Conference, Isfahan, May 1997.

7 See M. Stopford, "Shipping Risks and Risky Ships: Thoughts on the Approach to Market Forecasting for Tankers," *OPEC Review*

19 (2) (Summer 1995); Z.S. Zannetos, "Oil Tanker Markets: Continuity amidst Change," in *Energy: Markets and Regulation*, edited by R.L. Gordon, H.D. Jacoby and M.B. Zimmerman (Cambridge MA, MIT Press, 1987).

8 See Jacobs Holdings, *World Tanker Fleet Review* (London, Jacobs Holdings plc, 1995).

9 D. Yergin, *The Prize* (New York, Simon and Schuster, 1991).

10 Profit and loss can be complex. For example, in the 1960s, most refineries in Western Europe were posting losses, and yet refinery-building in this period in Western Europe was booming. The explanation for this apparent contradiction is that the oil companies were using their control over transfer prices from their crude-producing affiliates to switch profits to the more tax-benign producing countries. See L. Turner, *Oil Companies in the International System* (London, Royal Institute of International Affairs, George Allen and Unwin, 1980).

11 This was the norm until the 1960s, except in a limited number of cases. In these cases, crude long majors (i.e. those with access to more equity crude than refining capacity) did sell quantities to crude short majors on long-term contracts. In the 1960s, the entry of new companies upstream, many of whom had their crude stranded abroad by more stringent import requirements into the US, made more crude available outside of interaffiliate transfers.

12 See J.E. Hartshorn, "From Multinational to National Oil: The Structural Change," *Journal of Energy and Development* vol. 6 (Spring 1980); J.H. Mohnfeld, "Implications of Structural Change," *Petroleum Economist*, July 1982.

13 W. Hutton, *The State We Are In* (London, Jonathan Cape, 1995).

14 See M. Lynch, "Shoulder Against Shoulder: The Evolution of Oil Industry Strategy," *Journal of Energy and Development* vol. 19, no. 1 (1995); M. Ollinger, "The Limits of Growth of the Multi-Divisional Firm: A Case Study of the US Oil Industry from 1930-90," *Strategic Management Journal* vol. 15 (1994).

15 The distinction between producers and consumers is indistinct.

16 See D. Swann, *The Retreat of the State* (New York, Harvester Wheatsheaf, 1988).
17 See D. Upton, *Waves of Fortune* (Chichester, Wiley & Sons, 1996).
18 See F.E. Banks, "Oil Stocks and Prices," *OPEC Review* 18 (2) (Summer 1994); J. Roeber, *The Evolution of Oil Markets: Trading Instruments and their Role in Oil Price Formation* (London, Royal Institute of International Affairs, 1993); P. Verleger, "The Evolution of Oil as a Commodity," in *Energy: Markets and Regulation*, edited by R.L. Gordon, H.D. Jacoby and M.B. Zimmerman (Cambridge MA, MIT Press, 1987); P. Stevens, "Oil Prices: The Start of an Era?" *Energy Policy* vol. 24, no. 5 (1996).
19 See T. Lux, "Herd Behaviour, Bubbles and Crashes," *Economic Journal* 105 (431) (1995).
20 See G.C. Watkins & A. Plourde, "How Volatile are Crude Oil Prices?" *OPEC Review* 18 (4) (Winter 1994).
21 See M.A. Adelman, *The Economics of Petroleum Supply: Papers by M.A. Adelman 1962–1993* (Cambridge MA, MIT Press, 1993); D.G. Sauer, "Measuring Economic Markets for Imported Crude Oil," *The Energy Journal* vol. 15, no. 2 (1994).
22 M. Varzi, "World Energy Supply and Demand: Mid-Term Prospects for the Middle East," Middle East Oil & Gas: Towards the Third Millennium Conference, Isfahan, May 1997; H.T. Franssen, "Energy Policy Options for the Middle East," Middle East Oil & Gas: Towards the Third Millennium Conference, Isfahan, May 1997.
23 M. Grubb, "The Rise of Climate Change," in *The New Geopolitics of Energy*, edited by J.V. Mitchell et al., (London, Royal Institute of International Affairs, 1996).
24 See BP (Annual), *British Petroleum Statistical Review of World Energy* (London, British Petroleum, 1997).

1: Vertical Integration as a Strategy for Oil Security

1 See G. Luciani, *The Oil Companies and the Arab World* (London, Croom Helm, 1984), Chapter 7.
2 See G. Luciani, "The Dynamics of Reintegration in the International Petroleum Industry," in *Oil in the New World Order*, edited by K. Gillespie and H.C. Moore (Gainesville FL, University Press of Florida, 1995).
3 *Petroleum Intelligence Weekly*'s December special supplement issue on "Ranking of the World's 50 Largest Oil Companies," December 1996.
4 *Oil and Gas Journal*, December 18, 1995, p. 42.
5 See Salomon Brothers' "Weekly Domestic Regional Analysis of Margins" (DREAM) and "Margin Analysis of Global International Countries" (MAGIC) 1996 (New York, Salomon Brothers, 1996).
6 See R. Mabro and P. Horsnell, *Oil Markets and Prices: The Brent Market and the Formation of World Oil Prices* (Oxford, Oxford University Press, 1993).

2: International Downstream Integration of National Oil Companies

1 On the role of the industry structure in the evolution of Brent forward and futures market, see Paul Horsnel and Robert Mabro, *Oil Markets and Prices: The Brent Market and the Formation of World Oil Prices* (Oxford, Oxford University Press, 1993), pp. 73–83.
2 Between 1973 and 1985, oil's share in primary energy requirements of OECD nations declined from 54 percent to 40 percent. OPEC's share in the market declined from 55 percent to 29 percent. The effect of all this on OPEC's market power and its cohesion led to the oil price crash of 1986, which resulted in the abandonment of fixed prices and an increase in OPEC's share to 41 percent in 1995.

3 In the early 1980s, a variety of important acquisitions were concluded in the US oil industry, such as the Dupont/Conoco (US$4.4 bn), the Shell/Marathon (US$6.5bn), Texaco/Getty (US$10.1bn), Chevron/Gulf Oil (US$13.2bn), and Mobil/Superior (US$5.3bn).

4 Most of the deals involving NOCs made during the period were in the downstream, except for KPC's purchase of Santa Fe and its other ventures into upstream (See *Arab Oil and Gas Directory*, 1996).

5 The ratios for the NOCs are derived by taking refining capacities inside the respective countries (totally owned and joint ventures) taken from OPEC's *Statistical Bulletin* plus the equity shares of capacities overseas, taken from Table 2.1, divided by production of liquid oils by the NOCs. For the international majors, the ratios are derived from the figures in Table 2.3.

6 *Petrostrategies*, May 19, 1995.

7 Pierre Terzian, "Arguments For and Against Foreign Downstream Integration of Oil Exporting Countries," unpublished manuscript, (Paris, 1989).

8 Philip K. Verleger Jr., *Adjusting to Volatile Energy Prices* (Washington DC, Institute of International Economics, 1993).

9 John E. Treat, John Horbison, and Peter Pekar, "The Age of Collaboration," *Petroleum Intelligence Weekly*, September 30, 1996.

10 Clement B. Malin, "The Implications of Downstream Reintegration in Major Markets," Arab Refining Industry in the Nineties seminar, March 1989, OAPEC, Kuwait.

11 Edward Krapels, "Neo-Integration: Stabilizing or Disruptive," *Middle East Economic Survey*, August 12, 1991.

12 *Petroleum Intelligence Weekly*, December 18, 1995.

13 The IEA (1995) projects world oil demand to grow by 2 percent annually until the year 2010, with OECD demand growth at 1.2 percent and the developing countries at 4.2 percent. The World Bank (1995) projects world oil demand growth at 1.6 percent with the OECD's growth rate at 0.6 percent and the developing countries at 3.2 percent annually. OPEC (1996), on the other hand, projects world-oil-demand growth rate at 1.6 percent annually until 2010,

with 0.5 percent for OECD and 4.0 percent for the developing countries. The range of incremental oil demand forecast in the three models is approximately 1.3–1.8mb/d annually until 2010.

14 Philip K. Verleger, Jr., *Adjusting to Volatile Energy Prices* (Washington DC, Institute of International Economics, 1993), p. 183.

3: Strategic Options and Opportunities for Gulf Oil Companies in the Asian Market

1 Fereidun Fesharaki et al., *Pacific Energy Outlook: Strategy and Policy Imperatives to 2010* (Honolulu HI, East-West Center, 1995).

2 Jose Goldemberg, "Leapfrogging: A New Energy Policy for Developing Countries," *World Energy Council Journal*, December 1991.

3 Yasuyuki Nakabayashi and Kinji Ogawa, "Coal-Fired Power Generation for the Future Environment," *New Electricity* 21 (1996); Power Industry Technology and Management Strategies for the Twenty-First Century Conference, Tokyo, May 1992, published as *Conference Proceedings* (Paris, International Energy Agency, Organization for Economic Co-operation and Development, 1996).

4 Chantorn Saubhayana, "Trade and Transportation of Crude Oil and Petroleum Products," Symposium on Pacific Energy Cooperation 1996, Tokyo, February 1996.

5 Yong Ho Jang and Takashi Kato, 1996, "Ajia no Genyu Pricing to Kongono Kadai" (in Japanese) ("Crude Oil Pricing in Asia and its Future Issues"), *Energy in Japan* (Tokyo, Institute of Energy Economics, 1996).

4: Oil Refining and Petroleum-Product Specifications in Asia

1 M.A. Adelman asserted some time ago that "[the] world oil market, like the world ocean, is one great pool." See Adelman, "International Oil Agreements," *The Energy Journal* 5(3) (1984), pp. 1–9. A recent study by Gulen provides empirical evidence to support this contention, concluding that "the results strongly indicate that the world oil market is unified." See S.G. Gulen, "Regionalization in the World Oil Market," *The Energy Journal* 18(2) (1997), pp. 109–126.

2 *World Energy Outlook*, 1996, p. 237.

3 Ibid., p. 237.

4 The IEA projection for 2010 in Table 4.1 is based on the "capacity constraints" case, which assumes historical trends in energy efficiency improvements and an oil price at US$17/bl in real 1993 US dollar terms until 2000, rising to US$25/bl between 2000 and 2005, where it remains until 2010.

5 See, for instance, F. Fesharaki, H.S. Keevil, and J.P. Dorian, "Recent Trends in Domestic Energy Prices of Asia: Impacts of Deregulation and Changing Policy Directions," *OPEC Review* 20(2) (1996), pp. 117-148.

6 See *Petroleum Intelligence Weekly*, September 4, 1995.

7 See, for instance, *Weekly Petroleum Argus*, November 27, 1995; *Petroleum Intelligence Weekly*, October 30, 1995.

8 These were the available figures at the time of writing.

9 See *Weekly Petroleum Argus*, November 13, 1995, p. 4.

10 See Gilbert Wawoe, "The Outlook for the Refinery Industry in the Pacific Rim," Oil and Money conference, May 1993.

11 See, for instance, Hoesung Lee and Dennis Eklof, "Refining in the Far East: Its Potential and Constraints," *The Energy Journal*, Special Issue, 1994.

5: "Back-door" Privatization Initiatives in the Gulf: An Inadequate Step in the Right Direction

1 Wael Al-Mazeedi, "Privatizing National Oil Companies in the Gulf," *Energy Policy*, October 1992, pp. 983–994.

2 Gulf NOC's continue to be reluctant to privatize their core upstream and downstream businesses. Among the few examples of privatization in the Gulf is the Saudi Basic Industries Corporation (SABIC), 35 percent of whose shares are in public hands. The Saudi government is contemplating selling additional SABIC shares to the public. On a much smaller scale, Kuwait Petroleum Corporation (KPC) is considering the privatization of the 70 retail distribution outlets in Kuwait, currently operated by its refining subsidiary. However, disagreements concerning labor and antitrust issues continue to inhibit this privatization.

3 The concept of comparative advantage originated with David Ricardo, *Principles of Political Economy and Taxation* (London, Prometheus Books, 1996). The absolute advantage of Gulf hydrocarbon production has been demonstrated most recently by Thomas R. Stauffer, *Indicators of Crude-Oil Production Costs: The Gulf Versus Non-OPEC Sources* (Boulder CO, IRCEED, 1993).

4 Michael Porter, *The Competitive Advantage of Nations* (New York, Free Press, Simon & Schuster, 1990), argues that "competitive advantage" in international trade comes from the interaction of four country- and firm-specific elements:

1) factor conditions;
2) demand conditions;
3) related and supporting industry; and
4) firm strategy, infrastructure and rivalry.

Within Porter's framework, the "competitive advantage" of the Gulf is almost exclusively concentrated in the hydrocarbon industry, since its competitive position is derived solely from natural resources ("factor conditions"). This is, in Porter's framework, the most unsophisticated form of competitive advantage.

5 Vertical integration is the combination in a single firm of petroleum exploration and production ("upstream" activities), transportation, refining, and product distribution ("downstream" activities). See Stephen Kobrin, "An Empirical Analysis of the Determinants of Global Integration," *Strategic Management Journal* vol. 12 (1991). See also Peter Siddall, Keith Willey, and Jorge Taveres, "Building a Transnational Organization for British Petroleum," *Long Range Planning* vol. 25, no. 1 (1992).

6 To date, outsourcing has been the only significant conduit for private-sector participation in the hydrocarbon industry in many Gulf countries. Downstream services such as refinery and plant maintenance and catalyst handling are often outsourced. Upstream services such as contract drilling, field maintenance, wire-line logging, cementing, casing, testing, workovers, transportation, catering, and cleaning are outsourced. Recently, KPC entered into agreements with Chevron, BP, and Shell for the provision of technical services covering production development, reservoir management, and drilling. Similarly, KNPC entered into a technical services agreement with Conoco for its refineries.

7 One area of particular concern is the accounting relationship between the NOC and the Ministry of Electricity and Water. This interface is so inefficient that it often inhibits independent power-project development.

8 Traditionally, projects requiring large amounts of capital were financed by loans taken and backed by the developers. Consequently, the institutions that built infrastructure projects were those willing and able to undertake the associated political and commercial risks. Project financing, on the other hand, is a method of funding capital-intensive projects not based on the credit of any single entity, but on the revenues to be produced by the project itself. The project assets serve as the only collateral. The lenders look to the project's cash flow for repayment of principal and interest, just as equity investors look to the project for return on their investment. In its purest form, project financing is non-recourse lending based solely on the project.

Sponsors are not liable if the revenues of the project prove insufficient to meet the debt service commitment. In practice, project financing is better characterized as limited-recourse lending, primarily because the lenders may require some credit support or guarantees from the sponsors, the sponsors' affiliates, or other third parties such as governments, state-owned enterprises, multilateral development agencies, export credit agencies, or similar organizations, to cover some of the risks. Also, lenders typically require that the sponsors be entities with sufficient capital resources and track records to absorb risks of construction non-completion, cost over-runs, or sometimes even outright disruption of service.

9 Annalee Saxenian, *Regional Advantage: Culture and Competition in Silicon Valley and Route 128* (Cambridge MA, Harvard University Press, 1996).

10 Since OOC has relinquished its equity interest in the project, this example is provided only to illustrate the potential for downstream joint-venture integration by companies based in the Gulf region.

11 International energy companies are diversifying into independent power development on a global basis. Private oil companies are leveraging their knowledge of power generation in their own refining complexes. Gas companies see power generation as a natural fit, especially as combined-cycle gas turbines have become a highly efficient means for power generation.

Bibliography

Journal Articles and Conference Papers

"50 Years of Gas in Russia: A Special Supplement," *Petroleum Economist* vol. 63, no. 5 (May 1996).

"Asia's Downstream Expansion," *Oil & Gas Journal* vol. 95, no. 27 (July 7, 1997).

"Natural Gas in the Middle East," *Oil & Gas Journal* vol. 95, no. 8 (February 24, 1997).

"The Caspian Sea: Attracting Western Investment," *Petroleum Economist* vol. 64, no. 1 (January 1997).

"World Gas Conference Special Issue," *Petroleum Economist* vol. 64, no. 5 (May 1997).

"Mid-East Investment Gate Eases Open," *Euromoney* (London), supplement, November 1995).

Abi-Aad, Naji, and Michel Grenon, "Middle Eastern Geopolitics: Factors of Instability and Sources of Conflict," *Peace and the Sciences*, no. 26 (June 1995).

Adelman, M.A., "International Oil Agreements," *The Energy Journal* 5 (3) (1984).

Al-Mazeedi, Wael, "Privatizing National Oil Companies in the Gulf," *Energy Policy*, October 1992.

Altunisik, Meliha Benli, "Political and Security Implications of Oil in the Middle East," *Dis Politika/Foreign Policy* vol. 19, nos 3-4 (1995).

Amirahmadi, Hooshang, "World Oil and Geopolitics to the Year 2010," *Journal of Energy and Development* vol. 21, no. 1 (Autumn 1995).

— "Oil at the Turn of the Twenty-First Century: Interplay of Market Forces and Politics," *The Emirates Occasional Papers* no. 5 (1996).

Ash, Toby, et al., "Special Report: Petrochemicals," *MEED* vol. 41, no. 17 (April 25, 1997).

— "Special Report: Oil and Gas," *MEED* vol. 41, no. 15 (April 11, 1997).

— "Special Report: Power," *MEED* vol. 41, no. 32 (August 8, 1997).

Balabanoff, Stefan, "OPEC's View of the Oil Market in the Medium and Long Term," *OPEC Bulletin* vol. 28, no. 7 (July 1997).

Bamber, Derek, "Gas in Asia," *Petroleum Economist* vol. 63, no. 11 (November 1996).

Banks, F.E., "Oil States and Prices," *OPEC Review* 18, (2) (Summer 1994).

Barylski, Robert V., "Russia, the West and the Caspian Energy Hub," *Middle East Journal* vol. 49 (Spring 1995).

Berg, Andrew, and Elliot Berg, "Methods of Privatization," *Journal of International Affairs* vol. 50, no. 2 (Winter 1997).

Blank, Stephen, "Energy, Economics and Security in Central Asia: Russia and its Rivals," *Central Asian Survey* vol. 14, no. 3 (1995).

Calder, Kent E., "Asia's Empty Tank," *Foreign Affairs* vol. 75, no. 2 (March-April 1996).

Cohen, Ariel, "The 'New Great Game': Pipeline Politics in Eurasia," *Eurasian Studies* vol. 3, no. 1 (Spring 1996).

Colitti, Marcello, "The Need for Reintegration of the World Oil Industry," Third CGES Conference, London, April 1993.

Croissant, Michael P., "Oil and Russian Imperialism in the Transcaucasus," *Eurasian Studies* vol. 3, no. 1 (Spring 1996).

Dion, Richard R., "Central Asian FSU Report, Part 2," *Oil & Gas Journal* vol. 95, no. 30 (July 28, 1997).

Faroohar, Kambiz, "Iran: Big Oil Versus Clinton," *Middle East*, April 1995.

Feshakari, F. et al., "Recent Trends in Domestic Energy Prices of Asia: Impacts of Deregulation and Changing Policy Directions," *OPEC Review* 20 (2) 1996.

Feuilherade, Peter, "Iraq: Opening up the Oilfields," *Middle East*, April 1995.

Frank, C. Alexander, Central Asian FSU Report, Part 1, *Oil & Gas Journal* vol. 95, no. 29 (21 July 1997).

Franssen, H.T., "Energy Policy Options for the Middle East," Middle East Oil & Gas: Towards the Third Millennium Conference, Isfahan, May 1997.

Ghanem, Shokri, "High Stakes: How Climate Change Negotiations Could Affect Oil-Producing Countries," *OPEC Bulletin* vol. 28, no. 5 (May 1997).

Ghorban, Narsi, "The Evaluation of Recent Gas Export Pipeline Proposals in the Middle East," *Iranian Journal of International Affairs* vol. 7, no. 2 (Summer 1995).

Goldemburg, Jose, "Leapfrogging: A New Energy Policy for Developing Countries," *World Energy Council Journal*, December 1991.

Greenfield, Sarah, "The Asian Energy Industry Begins to Branch Out," *Petroleum Economist* vol. 62, no. 9 (September 1995).

Gulen, S.G., "Regionalization in the World Oil Market," *The Energy Journal* 18 (2) (1997).

Hamm, K., "Downstream Business: Economics and Future Prospects," Middle East Oil & Gas: Towards the Third Millennium Conference, Isfahan, May 1997.

Hartshorn, J.E., "From Multinational to National Oil: The Structural Change," *Journal of Energy and Development* vol. 5 (Spring 1980).

Hess, Andrew C., "Peace and Political Reform in the Gulf: The Private Sector," *Journal of International Affairs* vol. 49, no. 1 (Summer 1995).

Hindley, Angus, "Meeting the Gas Challenge," *MEED* vol. 40 (February 16, 1996).

Kalicki, Jan H., "A Vision for the US–Saudi and US–Gulf Commercial Relationship," *Middle East Policy* vol. 5, no. 2 (May 1997).

Kay, Jay G., "Oil Prices Go Soft," *Gulf Business* vol. 1, no. 12 (April 1997).

Khadduri, Walid, "Oil and Politics in the Middle East," *Security Dialogue* vol. 27, no. 2 (June 1996).

Knight, John, and Anthony Ling, "Mergers and Acquisitions," *Petroleum Economist* vol. 63, no. 12 (December 1996).

Knott, David, "Foreign Partners Scarce in OPEC Gulf States," *Oil & Gas Journal* vol. 94, no. 32 (August 5, 1996).

Kobrin, Stephen, "An Empirical Analysis of the Determinants of Global Integration," *Strategic Management Journal* vol. 12 (1991).

Krapels, Edward, "Neo-Integration: Stabilizing or Disruptive," *Middle East Economic Survey*, August 12, 1991.

Kutschera, Chris, "Azerbaijan: The Kuwait of the Caucasus?" *Middle East*, March 1996.

Lee, Hoesung, and Dennis Eklof, "Refining in the Far East: Its Potential and Constraints," *The Energy Journal*, Special Issue, 1994.

Lenczowski, George, "The Caspian Oil and Gas Basin: A New Source of Wealth?" *Middle East Policy* vol. 5, no. 1 (January 1997).

— "Major Pipelines in the Middle East," *Middle East Policy* vol. 3 (April 1995).

Lux, T., "Herd Behaviour, Bubbles and Crashes," *Economic Journal* vol. 105, no. 431 (1995).

Lynch, M., "Shoulder Against Shoulder: The Evolution of Oil Industry Strategy," *Journal of Energy and Development* vol. 19, no. 1 (1995).

Maddock, R.T., "Environmental Security in East Asia," *Contemporary Southeast Asia* vol. 17, no. 1 (June 1995).

Malin, Clement B., "The Implications of Downstream Reintegration in Major Markets," Arab Refining Industry in the Nineties seminar, March 1989, OAPEC, Kuwait.

Mattair, Thomas R., "Horizons for Cooperation in the Gulf and the Oman Sea: The View from Washington," *Iranian Journal of International Affairs* vol. 7, no. 3 (Fall 1995).

McLeod, Gordon, "Kyrgyzstan Starts up its First Refinery," *Oil & Gas Journal* vol. 95, no. 27 (July 7, 1997).

Mohnfeld, J.H., "Implications of Structural Change," *Petroleum Economist*, July 1982.

Nakabayashi, Yasuyuki and Kinji Ogawa, "Coal-Fired Power Generation for the Future Environment," *New Electricity* 21 (1996).

Penaloza, Humberto, "Petróleos de Venzuela's Experience in Joint Venture Downstream Arrangements," *Geopolitics of Energy*, October 1988.

Project & Trade Finance, February 1997.

Petroleum Intelligence Weekly, "Ranking of the World's 50 Largest Oil Companies," Special Supplement December 1996.

Ibid., September 4, 1995.

Ibid., October 30, 1995.

Ibid., December 18, 1995.

Petrostategies May 14, 1994.

Ibid., May 19, 1995.

Paik, Keun Wook, "Energy Cooperation in Sino-Russian Relations: the Importance of Oil and Gas," *Pacific Review* vol. 9, no. 1 (1996).

— "China, Oil and the Risk of Regional Conflict," *Survival* vol. 37, no. 4 (Winter 1995–1996).

Salomon Brothers (New York), *Weekly Domestic Regional Analysis of Margins* (*DREAM*), 1996.

— *Margin Analysis of Global International Countries* (*MAGIC*), 1996.

Saubhayana, Chantorn, "Trade and Transportation of Crude Oil and Petroleum Products," Symposium on Pacific Energy Cooperation 1996, Tokyo, February 1996.

Sauer, D.G., "Measuring Economic Markets for Imported Crude Oil," *The Energy Journal* vol. 15, no. 2 (1994).

Siddall, Peter, Keith Willey, and Jorge Taveres, "Oil Prices: The Start of an Era?" *Energy Policy* vol. 24, no. 5 (1996).

— "The Determination of Oil Prices 1945-95," *Energy Policy* vol. 23, no. 10 (1995).

— "Building a Transnational Organization for British Petroleum," *Long Range Planning* vol. 25, no. 1 (1992).

Stauffer, Tom, "Iran: Becoming a Hub for Central Asian Oil Exports," *Petroleum Economist* vol. 64, no. 7 (July 1997).

Stevens, Paul, "Increasing Global Dependence on Gulf Oil: 'This Year, Next Year, Sometime, Never?'" *Energy Policy* vol. 25, no. 2 (February 1997).

— "Oil Prices: The Start of an Era?" *Energy Policy* vol. 24, no. 5 (1996).

— "The Determination of Oil Prices 1945-95," *Energy Policy* vol. 23, no. 10 (1995).

Stopford, M., "Shipping Risks and Risky Ships: Thoughts on the Approach to Market Forecasting for Tankers," *OPEC Review* 19, (2) (Summer 1995).

Szabo, Antonio, M., "The Vertical Reintegration of the World Oil Industry," *MEES* 18 (July 1988).

Treat, John, John Horbison, and Peter Pekar, "The Age of Collaboration," *Petroleum Intelligence Weekly*, September 30, 1996.

Valencia, Mark J., "Energy and Insecurity in Asia," *Survival* vol. 39, no. 3 (Autumn 1997).

Varzi, M., "World Energy Supply and Demand: Mid-Term Prospects for the Middle East," Middle East Oil & Gas: Towards the Third Millennium Conference, Isfahan, May 1997.

Voge, B., Latham & Watkins, *Project Finance International*, January 29, 1997.

Watkins G.C., and A. Plourde, "How Volatile are Crude Oil Prices?" *OPEC Review* 18 (4) (Winter 1994).

Weekly Petroleum Argus, November 13, 1995.

Ibid., November 27, 1995.

WEFA Group, *Asia Economic Outlook*, February 1997.

WEFA Group, *Eurasia Economic Outlook*, February 1997.

WEFA Group, *Middle East and Africa Economic Outlook*, February 1997.

WEFA Group, *Asian Monthly Monitor*, March 1997.

World Bank, *Review and Outlook for the World Oil Market*, Discussion Paper no. 301, Washington DC (1995).

Zarsky, Lyuba, "The Prospects for Environmental Cooperation in Northeast Asia", *Asian Perspective* vol. 19, no. 2 (Fall–Winter 1995).

Books

Abi-Aad, Naji, and Michel Grenon, *Instability and Conflict in the Middle East: People, Petroleum and Security Threats* (Basingstoke, Macmillan, 1997).

Adelman, M.A., *The Economics of Petroleum Supply: Papers by M.A. Adelman 1962-1993* (Cambridge MA, MIT Press, 1993).

— *The World Petroleum Market* (Baltimore MD, John Hopkins University Press, 1972).

— *Genie Out of the Bottle: World Oil Since 1970* (Cambridge MA, MIT Press, 1995).

Agrawal, Pradeep et al., *Economic Restructuring in East Asia and India: Perspectives on Policy Reform* (Basingstoke, Macmillan, 1995).

Allan, J.A., and Chibli Mallat, *Water in the Middle East: Legal, Political, and Commercial Implications* (London, I.B. Tauris, 1995).

Arab Oil and Gas Directory (Paris, APRC, 1996).

Ayubi, Nazih N.M., *Distant Neighbours: the Political Economy of Relations Between Europe and the Middle East/North Africa* (Reading, Ithaca Press, 1995).

Bankoff, Greg, and Kylie Elston, *Environmental Regulation in Malaysia and Singapore* (Nedlands, University of Western Australia Press, 1995).

Beck, Robert J., *Worldwide Petroleum Industry Outlook, 1998-2002* (Tulsa OK, Pennwell, 1997).

Bettignies, Henri Claude de (ed.) *The Changing Business Environment in the Asia-Pacific Region* (London, Boston MA, International Thomson Business Press, 1996).

Blair, J., *The Control of Oil*, (London, Macmillan, 1976).

Bohi, Douglas R., and Michael A. Toman, *The Economics of Energy Security* (Boston MA, Kluwer Academic Publishers, 1996).

Booz Allen & Hamilton, *Petroleum Intelligence Weekly, Ranking The World's Top Oil Companies* (New York, Booz Allen & Hamilton, 1996).

BP (British Petroleum), *BP Statistical Review of World Energy 1996* (London, British Petroleum, 1996).

Carraro, Carlo, *International Environmental Negotiations: Strategic Policy Issues* (Cheltenham, Brookfield VT, Edward Elgar, 1997).

Chaudhry, Kiren Aziz, *The Price of Wealth: Economies and Institutions in the Middle East* (Ithaca NY, Cornell University Press, 1997).

CIA, *International Energy Statistical Review* (Washington DC, Central Agency, 1997).

Cook, Paul, and C.H. Kirkpatrick, *Privatisation Policy and Performance: International Perspectives* (New York, Prentice Hall/Harvester Wheatsheaf, 1995).

Cordesman, Anthony H., *Bahrain, Oman, Qatar, and the UAE: Challenges of Security* (Boulder CO, Westview Press, 1997).

— *Kuwait: Recovery and Security After the Gulf War* (Boulder CO, Westview Press, 1997).

Crystal, Jill, *Oil and Politics in the Gulf: Rulers and Merchants in Kuwait and Qatar* (Cambridge, Cambridge University Press, 1995).

Deegan, Heather, *Third Worlds: The Politics of the Middle East and Africa* (London, Routledge, 1996).

ECSSR, *Gulf Energy and the World: Challenges and Threats* (London, I.B. Tauris, 1997).

El-Erian, Mohamed A., *Growth and Stability in the Middle East and North Africa* (Washington DC, International Monetary Fund, 1996).

Feldman, David Lewis (ed.) *The Energy Crisis: Unresolved Issues and Enduring Legacies* (Baltimore MD, John Hopkins University Press, 1996).

Fesharaki, Fereidun et al., *Pacific Energy Outlook: Strategy and Policy Imperatives to 2010* (Honolulu HI, East-West Center, 1995).

Field, Graham, *Economic Growth and Political Change in Asia* (London, Macmillan, 1995).

Forsythe, Rosemarie, *The Politics of Oil in the Caucasus and Central Asia* (Oxford, Oxford University Press, 1996).

Fowler, David, and Gwilym Roberts, *Built by Oil* (Reading, Ithaca Press, 1995).

Fujime, K., "Long-Term Energy Supply/Demand Outlook for Asia APEC Nations" in *Energy in Japan* (Tokyo, Institute of Energy Economics, 1996).

Gilbar, Gad, *The Middle East Oil Decade and Beyond* (London, Frank Cass, 1996).

Grubb, M., "The Rise of Climate Change" in *The New Geopolitics of Energy*, edited by J.V. Mitchell et al. (London, Royal Institute of International Affairs, 1996).

Gleason, Gregory, *The Central Asian States: Discovering Independence* (Boulder CO, Westview Press, 1997).

Gordon, R.L., H.D. Jacoby, and M.B. Zimmerman (eds) *Energy: Markets and Regulation* (Cambridge MA, MIT Press, 1987).

Guazzone, Laura, *The Middle East in Global Change: The Politics and Economics of Interdependence Versus Fragmentation* (Basingstoke, Macmillan Press, 1997).

Halbach, Axel J., and H. Awartani, *New Potentials for Cooperation and Trade in the Middle East* (Munich, Weltforum Verlag, 1995).

Hartshorn, J.E., *Oil Trade: Politics and Prospects* (Cambridge, Cambridge University Press, 1993).

Henmi, Kenzo, and John C. Cool, *Educating New Environmental Leadership for Asia: The Report of the Initiative for Development of Environmental Alliances Through Leadership* (Arlington VA, Winrock Publications, 1995).

Henderson, Simon, *The Middle East in the Year 2000: New Opportunities, New Dangers* (London, FT Energy Publishing, 1996).

Horsnell, Paul, *Oil in Asia: Markets, Trading, Refining, and Deregulation* (Oxford, Oxford University Press, 1997).

Horsnell, Paul, and Robert Mabro, *Oil Markets and Prices: The Brent Market and the Formation of World Oil Prices* (Oxford, Oxford University Press, 1993).

Hutton, W., *The State We Are In* (London, Jonathan Cape, 1995).

Ikeo, Aiko, *Economic Development in Twentieth Century East Asia* (London, Routledge, 1997).

International Energy Agency, *Asia Gas Study* (Paris, OECD, 1996).

— *Middle East Oil and Gas* (Paris, OECD, 1995).

— *Oil Market Report: Statistical Supplement* (Paris, OECD, 1996).

— *Oil Supply Security: The Emergency Response Potential of IEA Countries* (Paris, OECD, 1995).

International Monetary Fund, *World Economic Outlook* (Washington DC, International Monetary Fund, 1997).

Jang, Yong Ho, and Takashi Kato, "Ajia no Genyu Pricing to Kongono Kadai" (in Japanese) ("Crude Oil Pricing in Asia and its Future Issues") (Tokyo, Institute of Energy Economics, 1996).

Jacobs Holdings, *World Tanker Fleet Review* (London, Jacobs Holdings plc, 1995).

Jha, Veena, Grant Hewison, and Maree Underhill, *Trade, Environment and Sustainable Development: A South Asian Perspective* (Basingstoke, Macmillan, 1997).

Kalland, Arne, and Gerard Persoon (eds) *Environmental Movements in Asia* (Surrey, Curzon Press, 1997).

King, Victor T. (ed.) *Environmental Challenges in South-East Asia* (Surrey, Curzon Press, 1997).

Larkin, Kay (ed.) *Financial Times, Oil and Gas International Year Book 1996* (London, Cartermill International, 1995).

Lu, G.Q., et al. (eds) *Sustainable Energy and Environmental Technology, Proceedings of the Asia-Pacific Conference*, Singapore, June 19-21, 1996 (Singapore, World Scientific Publishing, 1996).

Luciani, Giacomo, "The Dynamics of Reintegration in the International Petroleum Industry" in *Oil in the New World Order*, edited by K. Gillespie and H.C. Moore (Gainesville FL, University Press of Florida, 1995).

— *The Oil Companies and the Arab World* (London, Croom Helm, 1984).

Mendelsohn, Robert, and Daigee O. Shaw, *The Economics of Pollution Control in the Asia Pacific* (Cheltenham, Brookfield VT, Edward Elgar, 1996).

Mitchell, J.V. et al., *The New Geopolitics of Energy* (London, Royal Institute of International Affairs, 1996).

Naisbitt, John, *Megatrends Asia: Eight Asian Megatrends that are Reshaping our World* (New York, Simon & Schuster, 1996).

Nakabayashi, Yasuyuki, and Kinji Ogawa, "Coal-Fired Power Generation for the Future Environment," *New Electricity* 21: Power Industry Technology and Management Strategies for the Twenty-first Century Conference, May, 1992, Tokyo, published as *Conference Proceedings* (Paris, International Energy Agency, Organization for Economic Co-operation and Development, 1996).

Nawwab, Ismail I., Peter C. Speers, and Paul F. Hoye, *Saudi Aramco and its World: Arabia and the Middle East* (Dhahran, Saudi Arabia, Saudi Aramco, 1995).

Oil & Energy Journal, International Energy Statistics Sourcebook (Tulsa OK, Pennwell, 1996).

Organization for Economic Co-operation and Development, *Privatisation in Asia, Europe and Latin America* (Paris, OECD, 1996).

Organization of Arab Petroleum Exporting Countries, *Secretary-General's Twenty-First Annual Report* (AH 1414-1415/AD 1994, Kuwait, OAPEC, 1995).

Paik, Keun Wook, *Gas and Oil in Northeast Asia: Politics, Projects and Prospects* (London, Royal Institute of International Affairs, 1995).

Porter, M., *The Competitive Advantage of Nations* (New York, Free Press, Simon & Schuster, 1990).

Ramanadham, V.V., *Privatization and Equity* (London, Routledge, 1995).

Resources Institute, UNEP, UNDP, & World Bank, *World Resources 1996-97* (New York and Oxford, Oxford University Press, 1996).

Ricardo, David, *Principles of Political Economy and Taxation* (London, Prometheus Books, 1996).

Richards, Alan, and John Waterbury, *A Political Economy of the Middle East* (Boulder CO, Westview Press, 1996).

Roeber, J., *The Evolution of Oil Markets: Trading Instruments and Their Role in Oil Price Formation* (London, Royal Institute of International Affairs, 1993).

Sassanpour, Cyrus, *Policy Challenges in the Gulf Cooperation Council Countries* (Washington DC, Middle Eastern Department, International Monetary Fund, 1996).

Saxenian, Annalee, *Regional Advantage: Culture and Competition in Silicon Valley and Route 128* (Cambridge MA, Harvard University Press, 1996).

Seal, R., *Downstream Oil: 'De-integrated Profitability or Ruin?'* (London, British Petroleum, 1991).

Shafik, Nemat, *Claiming the Future: Choosing Prosperity in the Middle East and North Africa* (Washington DC, The World Bank, 1995).

Sharma, Shankar (ed.) *Energy, Environment and the Oil Market: An Asia-Pacific Perspective* (Singapore, Institute of Southeast Asian Studies, 1995).

Shirk, Susan L., and Christopher P. Twomey, *Power and Prosperity: Economics and Security Linkages in Asia-Pacific* (New Brunswick NJ, Transaction Publishers, 1996).

Shojai, Siamack, *The New Global Oil Market: Understanding Energy Issues in the World Economy* (Westport CT, Praeger, 1995).

Simon, Denis Fred (ed.) *Corporate Strategies in the Pacific Rim: Global vs Regional Trends* (London, Routledge, 1995).

Smith, Graham, et al., *Getting Connected: Private Participation in Infrastructure in the Middle East and North Africa* (Washington DC, World Bank, 1997).

Spiegel, Steven L, and David J. Pervin (eds) *Practical Peacemaking in the Middle East: The Environment, Water, Refugees, and Economic Cooperation and Development* (New York, Garland, 1995).

Stauffer, Thomas R., *Indicators of Crude-Oil Production Costs: The Gulf Versus Non-OPEC Sources* (Boulder CO, IRCEED, 1993).

Stevens, Paul, "A Survey of Structural Change in the International Oil Industry 1945-1984" in *The Changing Structure of the World Oil Industry*, edited by D. Hawdon (London, Croom Helm, 1984).

Swann, D., *The Retreat of the State* (New York, Harvester Wheatsheaf, 1988).

Tétreault, Mary Ann, *The Kuwait Petroleum Corporation and the Economics of the New World Order* (Westport CT, Quorom Books, 1995).

Tsuushousangyoushou Shigen-enerugi-chou, *Ajia Enerugi Bijon: Sougou Enerugi Chousakai Enerugi Bukai Chukan Houkoku* (in Japanese), (*Asian Energy Outlook: Interim Report of the International Division of the Advisory Committee for Energy*), (Tokyo, Tsuushousangyou Chosakai, 1995).

Turner, L., *Oil Companies in the International System* (London, George Allen and Unwin, 1980).

United States Senate Committee on Foreign Relations, *United States Dependence on Foreign Oil.* Hearing, March 27, 1995, 104th Congress, 1st Session, Senate Hearing 104–21 (Washington DC, Superintendent of Documents, 1995).

UNCTAD, *Trade and Development Report 1996* (New York, United Nations, 1996).

Upton, D., *Waves of Fortune* (Chichester, John Wiley & Sons, 1996).

Verleger, Philip K., *Adjusting to Volatile Energy Prices* (Washington DC, Institute of International Economics, 1993).

— "The Evolution of Oil as a Commodity" in *Energy: Markets and Regulation*, edited by R.L. Gordon, H.D. Jacoby & M.B. Zimmerman (Cambridge MA, MIT Press, 1987).

Wilson, Rodney, *Economic Development in the Middle East* (London, Routledge, 1995)

Yarrow, George K., and Piotr Jasinski, *Privatisation: Critical Perspectives on the World Economy* (London, Routledge, 1996).

Yeats, Alexander J., *Export Prospects of Middle Eastern Countries: A Post-Uruguay Round Analysis* (Washington DC, International Trade Division, The World Bank, 1996)

Yergin, Daniel, *The Prize* (New York, Simon and Schuster, 1991).

Zannetos, Z.S. "Oil Tanker Markets: Continuity Amidst Change" in *Energy: Markets and Regulation*, edited by R.L. Gordon, H.D. Jacoby & M.B. Zimmerman (Cambridge MA, MIT Press, 1987).

Index